napa bulletin 15

T0201472

Global Ecosystems: Creating Options through Anthropological Perspectives

■ Pamela J. Puntenney, ed.

National Association for the Practice of Anthropology
A section of the American Anthropological Association

NAPA Bulletins are occasional publications of the National Association for the Practice of Anthropology, a section of the American Anthropological Assocition.

Ralph J. Bishop and Pamela Amoss
General Editors

Library of Congress Cataloging-in-Publication Data

Global ecosystems: creating options through anthropological perspectives / Pamela J. Puntenney, ed.
    p.   cm.—(NAPA bulletin ; 15)
    Includes bibliographical references.
    ISBN 0-913167-70-3
    1. Environmental policy. 2. Economic development—Environmental aspects.
3. Sustainable development. 4. Applied anthropology.   I. Puntenney, Pamela
J. II. Series.
GE195.G56   1995
363.7—dc20   95-1455
                                                                          CIP

ISBN 0-913167-70-3

Contents

Forward 1

Preface 2

Solving the Environmental Equation: An Engaging Anthropology
   Pamela J. Puntenney 4

Alleviating Poverty and Conserving Wildlife in Africa:
An "Imperfect" Model from Zambia
   R. Michael Wright 19

Anthropology in Pursuit of Conservation and Development
   Michael Painter 33

Anthropology, the Environment, and the Third World:
Principles, Power, and Practice
   David D. Gow 46

The Tropical Forestry Action Plan: Is It Working?
   Robert Winterbottom 60

Ecological Awareness and Risk Perception in Brazil
   Alberto C. G. Costa, Conrad P. Kottak, Rosane M. Prado, and John Stiles . . . 71

The Cultural Environment of Development: Commentary
   Terence Turner 88

About the Contributors 98

Foreword

Global Ecosystems: Creating Options through Anthropological Perspectives

Roy A. Rappaport

It would be an exaggeration to claim that the future of anthropology lies entirely in its engagement with the troubles of the contemporary world, but it would be an equally, or even more, serious error to continue to relegate such engagement to anthropology's peripheries. The number of anthropologists devoting all, or a significant part, of their professional efforts to the understanding and amelioration of contemporary difficulties has grown in recent years and, judging from the interests of graduate students at my institution, will only increase further in the decades to come. There is no end of substantive matters, from substance abuse to ozone depletion, for them to engage. Furthermore, anthropological engagement with contemporary difficulties is a growth industry, whereas academic anthropology may not be.

I do not mean to make a sharp distinction between academic and nonacademic anthropologists or between their general projects, one "theoretical," the other "applied." That such a distinction has existed is unfortunate, and we should all do whatever we can to obliterate it. For one thing, there are important, perhaps even fundamental, theoretical payoffs for anthropology in dealing with the world's disorders—not the least of which may be in conceptualizing disorder itself. Anthropology, however, cannot engage the world's disorders if it is not allowed to bring its own conceptions to bear upon them. Anthropologists working on matters such as development have frequently done so from positions of weakness, subordinate to others, most often development economists, who have planned and administered the programs and whose simple-minded views of the world as "bottom lineable" usually prevail. Anthropological techniques, shorn of anthropological perspectives, are thus put into the service of ends that have been defined by neoclassical economics.

Regnant economic paradigms must be challenged, but challenging them is not enough. If an engaged anthropology is to be effective, we must offer paradigms of our own, paradigms that include not only the "microanthropology" of ethnography but also the "macroanthropology" of approaches such as world system theory, linkage theory, and theory of adaptive structures.

The interactions of culture, environment, and development form a strategic focus within the areas that must be addressed by a theoretically informed, engaged anthropology. This volume, as concerned with the conflicts and contradictions of working in this field as with the substantive problems of the field itself, makes an important contribution to the developing role of anthropology in environmental studies. I am confident that this thoughtful and useful collection of articles will help more of us to focus upon its concerns.

Preface

P. J. Puntenney

Two decades ago, natural resource management was primarily defined in terms of the economic value of the resource. Attempts to regulate and manage human environmental processes and solve environmental problems were primarily accomplished through legal, technical, and economic methods. Reflecting a worldview based upon the assumption that humans are separate from their environment, this approach to environmental problem solving allowed valued resources to be controlled by political and technical devices.

By the 1980s, humanity was beginning to confront its environmental problems more realistically. Advances in computer technology and satellite imagery increased the world's awareness of environmental issues and broadened our understanding of the magnitude of environmental change. By the early 1990s, the environmental outlook had shifted from a concern with problems impacting human health and welfare to processes defined in terms of interrelationships and the sustainability of environmental and human systems.

The human factor is apparent in directly addressing the ensuing conflict among local, regional, national, and international environmental and socioeconomic interests. To build practical policy considerations into the development agenda, the needs of the community-at-large must be defined in a useful form to avoid transferring inappropriate solutions. This kind of practical approach to policy making requires the participation of those affected. Whether cultural brokers or intermediaries, anthropologists are an important part of this formula. Bringing a cross-cultural perspective that draws from a number of disciplines and areas of professional expertise in gathering concrete data, anthropologists can define and identify an expanded repertoire of options. Anthropologists can be particularly useful in redefining and expanding potential roles and bringing people together in mutually beneficial ways, from the local initiative to the development of a national environmental plan of action.

We are beginning to move toward a proactive approach to sustaining the biosphere, recognizing that solutions must be specific to individuals, the community, places, and the conditions for economic development strategies that deal with regional and local realities. Michael Orbach, who received the 1991 Praxis Award from the Washington Association of Professional Anthropologists (WAPA), illustrates these trends through his pioneering work with the marine fisheries community. Using an analytical and applied framework, he bridged constituent and community involvement to shape state and influence federal environmental policy. His consideration of the "total" human system, from the fishers to the agencies and commissions, the legislatures, and the scientific community, defined the context and content for a new management and policy system to regulate the industry. More salient examples such as this are needed from the environmental community.

Although anthropology is increasingly making discernible contributions to the work on a "sustainable" biosphere, it is not clear how it fits into the equation. This NAPA Bulletin is the first of a series of bulletins focused on international environmental issues. The bulletin began as an American Anthropological Association Program Committee session in 1989 entitled "Culture, Environment, and Development: Interdisciplinary Perspectives on Critical Issues." Reflecting the current thinking about international environmental problem solving, the symposium brought together an interdisciplinary panel representing nonprofit, international development, and government organizations, policy institutions, and the academic community. This volume is an outgrowth of their discussions about creating policy, programs, and action-based alternatives that apply anthropological perspectives.

This bulletin articulates the role of the anthropological perspective in defining solutions adding to our knowledge of how to create appropriate institutional arrangements that are responsive to environmental issues. Each author has been involved in the creation of these new institutional structures. Taking stock of existing knowledge about the scope of the critical issues, these articles set forth recommendations for future efforts. In "Alleviating Poverty and Conserving Wildlife in Africa: An 'Imperfect' Model from Zambia," Michael Wright, African Wildlife Foundation, examines the interrelationships between wildlife management and human needs in Africa and concludes that participation is a critical variable in the decision-making process. In "Anthropology in Pursuit of Conservation and Development," Michael Painter, Institute of Development Anthropology, argues that the discipline of anthropology needs to learn what is known about the functions and limitations of natural systems and environmental degradation before it can make substantial contributions toward resolving environmental issues. In "Anthropology, the Environment, and the Third World: Principles, Power, and Practice," David Gow, World Resources Institute, concentrates on the role of the practicing anthropologist in the international environment-development arena, stressing the political context as essential for the practitioner's work and the necessity of communicating anthropological insights to diverse audiences. In "The Tropical Forestry Action Plan: Is it Working?", Robert Winterbottom, International Resources Group, Ltd., using the Tropical Forestry Action Plan as a case study, demonstrates the limitations of developing multinational and internationally initiated strategies without the benefit of input from the anthropological community. In "Ecological Awareness and Risk Perception in Brazil," Alberto Costa et al., University of Michigan, examine the issue of environmental awareness in the context of Brazil. They set the stage for developing substantive theory that is grounded through the use of a linkages methodology for understanding environmental risk and risk perception. In "Solving the Environmental Equation: An Engaging Anthropology," Terence Turner, University of Chicago, concludes with a commentary on the trend in anthropology toward an empowering, action-based model. These articles, in part, represent the new forms of emerging leadership within anthropology, forming specific recommendations for improving the policy-making process, the application of anthropological knowledge, research needs, and appropriate methodologies. They should be useful to professionals interested in public policy making and professionals interested in the importance of understanding the ecological and human processes of environmental change.

Articles

Solving the Environmental Equation: An Engaging Anthropology

Pamela J. Puntenney

Anthropologists engaged in environmental work are finding themselves in a challenging new era framed by the internationalization of environmental problems and policies associated with the globalization of economic considerations. Yet, paradoxically, there is an even greater need for national and local autonomy in defining environmental priorities. It has become clear to decision makers that both top-down and bottom-up approaches involving diverse groups of people are required to solve the environmental equation.

The disorders of the ecosphere induced by human activity indicate that something is fundamentally wrong with the cultural adaptations of modern society. Our local and global mindset shape our values, lifestyles, and practices, influencing all that we do no matter how we enhance the efficacy of our institutions or how we restructure legal and moral decision making. Without altering our basic values, we have consistently relied upon institutional efforts to solve problems through technical and technological change and behavioral solutions to modify actions of individual responses.

Questions arise out of the failures of the two strategies as we continue the search for viable options (Escobar 1991; Gow 1991). Institutional and individual responses are influenced by the historical, cultural, and economic forces that shape our modern worldview. Yet in order to move beyond this state of affairs, we need new cultural forms that conform to ecological realities. To achieve a common ground of understanding, we must incorporate diverse cultural perspectives into our thinking while keeping the ecosphere as an essential point of reference. As anthropologists learn how to confront the contradictions of mutual interdependence and relative autonomy within the context of ecological systems, they are increasingly finding themselves in positions where the utilization of anthropological perspectives has become essential to effective policy making, program development, and project implementation (Colson 1985; Hackenberg 1976; Johnson and Orbach 1990; Kottak 1990; Wulff and Fiske 1987).

This volume offers answers to the question: Is it possible for the human component of the Earth's ecosystems to live with its own ability to alter natural systems, or will humans continue to destroy the environment that sustains them?

Reconsidering the Global Village

Over the last three decades, environmental problems of pollution and degradation have changed not only in terms of scale and extent but also in their essential characteristics (Arizpe 1991; Gallopin 1991; Wilson 1988). It was not until the mid-

1980s, when advances in computer technology and satellite imagery were developed, that our understanding of the scale and complexity of environmental issues was broadened. In the 1990s we find that cause-and-effect relationships have become more difficult and complex both scientifically and legally. It is now clear that environmental issues are more closely associated with social and economic systems than technical solutions. In turn, we have come to realize that the kinds of problems we face cannot be resolved without anticipatory and preventative measures. Another change is the emergence of problems that are international in either origin or effects or both, such as acid rain, the transboundary movement of hazardous and toxic wastes, global warming, depletion of the ozone layer, and the disappearance of tropical rainforests. Achieving workable solutions to these types of problems requires shared resources and an ability to respond in a coordinated fashion from the international to the local levels (J. Mathews 1991).

It is important to note that international environmental problems impacted by economic activities have different kinds of causes and, consequently, differ in the actions needed to be taken. Akiyama (1993) suggests five types.

The first type, transboundary and regional environmental pollution (notably of air and international waterways), has traditionally been viewed as a serious problem at the local and national levels. However, the causes and effects from these kinds of transboundary pollution involve multiple nations settling legal problems and, on an international scale, require a broader consideration of economic costs and benefits.

The second type is environmental disruption caused by "pollution export" from foreign industries, often the result of direct investment by advanced countries with stringent environmental regulations into less-advanced countries with lower environmental standards. Dramatic examples are the Bhopal accident in India and the deals struck in West Africa to deposit toxic wastes there. The effects of this problem demand a response that is based upon a sociocultural system that enables each economic body to carry out its environmental responsibility.

The third type is environmental degradation caused by the international linkages of industry and trade as in the commercial harvesting of tropical rainforests and the over fishing of coastal waters. These environmental problems demonstrate the need to revise our international trade systems (e.g., GATT), develop an international industrial policy, and design a system of consumption that is environmentally sound.

The fourth type is the total collapse of a local or regional ecosystem brought about by the vicious cycle of poverty and environmental degradation. Examples are found in the new term of "environmental refugees" and by the massive flooding, soil erosion, and eutrophication of rivers caused by the deforestation of the Himalayan watershed. The challenge to the developed countries is how to create a new concept of international cooperation that would include economic development aid and environmental aid as a development strategy to the affected countries.

The last type is the contamination and degradation of the global commons such as the oceans, the stratosphere, and the atmosphere. Ozone depletion, the dumping of nuclear waste canisters into the oceans, and global warming are good examples. In this last category, the relation between cause and damage is not direct. Immediate effects cannot be observed at the local level. Yet successful solutions call for action at the local level. The present world system will have to develop appropriate policies and plans of action utilizing each nation's commitment. Unfortunately, this approach has had only limited results to date. The alternative is to devise mechanisms whereby a broad-based community in each nation can have input into the process.

As environment and development issues steadily become even more global in scope, the strict dividing line between foreign and domestic environmental policies is blurring. As demonstrated in Rio at the Earth Summit conference on environment and development, simple alignments of North-South or East-West are no longer sufficient. Elements associated with the one are also found within the other, such as poverty, population pressures, diverse cultures, technical capabilities, and scales of economy. We have come to the realization that solving environmental problems often requires broad cooperation and coordination of agreed-upon priorities among competing interests with contradictory aims. Although science can attempt to evaluate the functions and risks confronting ecological systems, solutions to the major environmental issues ultimately will require public choice and public responsibility (Engel and Engel 1990; Stapp 1984, 1986). Mechanisms are needed that draw upon multiple voices, especially from the people who will be directly impacted and will have to sustain an enduring level of commitment over an extensive period of time—perhaps several generations.

Heightened awareness of these challenges has brought about some degree of change in how we approach environmental problems. Ecosystems, however, are far from homogeneous and are often arbitrarily defined for purposes of scientific study or political expediency. We continue to artificially oppose sociocultural and natural systems, perceiving nature as outside the system waiting to be transformed into "rational-intentional forms . . . in the service of humanity." As a consequence, most considerations of the environment-development dilemma continue to focus either on socioeconomic concerns or on the dynamics of organisms within particular ecosystems, with little attention to their interconnectedness.

Within the context of these interrelationships the central questions are: What are the ways in which human society depends on the normal functioning of natural systems? How are solutions to be defined and by whom? How do social scientists transform knowledge into practice in order to bridge the social/cultural, political, economic, technical, and ecological dimensions of human lifestyles?

History and Process of Developing the New Wisdom

Given what we have learned about the ecosphere, we acknowledge that the continuing vitality of the social context cannot be insulated from the life-supporting processes of the world. Traditionally, however, the study of natural systems has been conducted within disciplinary fragments such as physics, chemistry, biology, sociology, anthropology, psychology, natural resources, and theology. And, in the tradition of good science, scholars have tried to synthesize these incommensurable results, post hoc, into views of the whole.

Similarly, policy makers have tended to frame environmental problems in terms of economic constructs, pollution, management of particular species, land-use degradation, technical and scientific solutions, and regulatory/legislative measures. As a society, we have conceived "nature" as splendidly isolated "wilderness," applying conservation strategies that have been exclusionary and misanthropic. We preserve this outdated notion by setting aside small refuges, parks, protected areas, buffer zones, and sanctuaries, usually inadequate in size, to support the ecosystems or organisms they are meant to protect (Gomez-Pompa and Kaus 1990; Wright 1993). Meanwhile, throughout the world, biodiversity is lost, ecosystems continue to be destroyed or degraded, and plant and animal species are still going extinct at an unprecedented rate. As a result, we are brought to recognize a more dynamic relationship with the realities of the ecosphere (Puntenney 1994).

Concerning the study of nature, anthropologist Edmund Leach shrewdly noted:

> The human observer stands apart; he is not personally involved. But this, of course, is just a fiction: in reality, the human observer and the stuff he observes share the same natural qualities, and this gives the whole business an uncomfortable air of relativity. The scientific study of nature is like Alice in Wonderland's game of croquet in which the mallets were flamingoes and the hoops kept walking off the ground. In this context, the scientist's insistence on detachment is simply an attempt to impose order on an unstable situation, a device to overcome the anxiety which arises from his inability to bring everything under human control. It is the modern substitute for prayer and primitive magic. [1968:11]

It is clear that if both economic development and human society are to be sustained, they must coexist within the limits of the Earth's natural systems (Soule 1986). Is it possible to have both development and maintenance of a quality environment? The authors in this volume respond with a qualified "yes" and consider how human activities can be managed so that we do not destroy the environment upon which we depend.

From Particulars to Patterns

A hundred years ago the concept of ecology did not yet exist, and humans were not seen as part of nature; by the second half of the 20th century, however, the fiction of human self-sufficiency had been dispelled. Though Tansley (1935) wrote about the earth's life-supporting processes and their interrelationships, the language eluded us until Lamont Cole (1958) conflated the terms "ecosystem" and "biosphere" into "ecosphere" to more accurately reflect the reality of a global entity.

By the late 1960s, scientists from around the world had become alarmed by evidence of impending environmental crisis and came to realize that studying the environment was not enough; the time had come for them to enter the political arena. In 1972, the Stockholm Conference on the Human Environment was organized for the purpose of bringing worldwide attention to environmental problems. Although the conference has been criticized for only reaching out within the United Nations community, it laid the groundwork for all that has followed in this arena: (1) scientists made an unprecedented decision to place the environment on the international political agenda; and (2) the convenors designed the international conference as a forum with a working agenda that would culminate in specific actions to be implemented within a designated timeframe.

By the mid-1970s environmental investigation had already shifted in focus from the "problems" themselves to their underlying causes. Some ten years later, the environment entered public discussion worldwide, influenced by the publicity about disasters in such places as Love Canal, Chernobyl, Bhopal, and Prince William Sound. Warnings regarding the global effects of the destruction of the Brazilian rainforests and growing piles of radioactive waste also captured public concern. And when, in 1988, James Lovelock's well-publicized Gaia theory identified the three-dimensional planetary skin as an integrated life-supporting system, our long-held view of the world as an aggregate of distinct units was further challenged (Lovelock 1990; Schneider 1989). These events not only raised public concern but also educated the public about the environment, placing pressure on both the scientific community and the policy makers to attend to global interrelationships.

As we near the 21st century, our understanding of the ecosphere has expanded to a growing recognition that "ecosystems are neither organisms nor super-organism." We have come to realize that ecosystems are different from and more important than populations, communities and organisms, the object of our traditional focus. There has been a movement away from a species-centered ethic and the mode of thought that serves it. Rowe (1961, 1989) helps us to understand that

> The Ecosphere is realistically conceived as comprising a hierarchy of ecosystems, like boxes within boxes, defined at various scales—zonal, regional, and local—for purposes of contemplation, study and ministration. These sectorial ecosystems—simplistically named seas, continents, mountains, plains, deserts, forests, lakes, rivers, settled lands,farm fields, towns, according to prominent natural or cultural features—posses an importance that far transcends their contents.
>
> The myriad forms of evolved life are the historic fruits and contemporary components of these evolved volumes. Humanity came into being within regional ecosystems—forest, savannah, grassland, seashore—as symbiotic parts of them, co-evolved with them, inseparable from them, along with a host of companion organisms of equal merit and importance.
>
> Living things arose within the ecosystems that the Ecosphere comprises. Thus the truth: Life is a phenomenon of the Ecosphere. Life is not something possessed by organisms, except in a limited and incomplete sense. From this corollary: "Ecosystems have organisms" is a more discerning idea than the conventional "Organisms have environments." [1989:125]

Today, we are beginning to understand that how we separate out and evaluate our many interventions in the cycles of the ecosphere and its ecosystems depends in part on our knowledge of the functioning of natural systems and in part on our relationship with renewable and replenishable components of these systems. At the most fundamental level, our understanding must include these constructs:

- *Renewable components of natural systems* are organisms that can increase in quantity in the ecosphere (microorganisms, plants, animals, and humans) through their reproductive capacities.
- *Replenishable components of natural systems* comprise the "matrix of ecosystems" (air, water, soil, and climate, including solar radiation) with their quantity and quality determining the level of support and the limits of productivity for renewable components of natural systems.
- *Nonrenewable components of natural systems* are finite sources of energy (fossil fuels—coal, oil, natural gas) within the Earth's crust used for purposes of economic development. The by-product is the release of waste residues and waste heat.
- *Nonreplenishable components of natural systems* are extracted from within the Earth's crust (subsurface ore bodies and industrial minerals) and are "potent sources of pollutants and toxins." Once they have been transformed they are for the most part no longer of use and often not recyclable, contributing to the growing pollution problem.

Our environmental problems in reality are "people/resource" issues shaped by (1) human demands that are in excess of the systems' ability to restore themselves and (2) the release of wastes from nonrenewable sources and pollutants from nonreplenishable sources that stress the health of regional and local ecosystems and, in turn, the ecosphere, thus affecting the systems' ability to renew and re-store a dynamic balance.

What does anthropology bring to the solution of our problem? Anthropology brings more than an obvious source of intercultural translations. The discipline of-fers a rich source of knowledge regarding viable forms of social production, cultural self-determination, ecological integration, subsistence resource base protection, policy formation, and defense of local autonomy. Perhaps the most important task at hand is to define ways in which we, as anthropologists, can clarify our under-standing of the dynamic interactions that are a part of larger systems. We must as-sist in the development of a language for broad-based communication concerning the ecosphere and environmental issues across national boundaries and among competing interest groups. We also need to develop models that will allow action on the internationalization of environmental policy and the process of public deci-sion making.

Conceptual Framework

Anthropologists trying to frame such a universal language and to formulate ap-propriate actions will need to consider the two interrelated concepts that are impor-tant to the study of human and environmental systems: sustainability, and the utili-zation of knowledge. The concept of sustainability encompasses scientific and economic paradigms that constitute current thought regarding "managing ecosys-tems." The dilemma of ecological systems versus economic systems has been criti-cized in environment and development literature, which calls for a broad concep-tualization that more accurately reflects the complexity and ambiguity of biological systems (Daly 1990). Within the research paradigms on ecological systems human systems components are mentioned but more often than not are omitted in both the conceptual and operational design of the research process. The human systems models likewise acknowledge the existence of natural systems' influences but commonly leave the fundamental constructs out of research and practice (Sponsel 1986). In recent critical discussions that debate the importance of fact versus value, means versus ends, and instrumental versus technical relationships, these compo-nents are now defined as relational rather than independent from each other (De-Walt 1988; McCay and Acheson 1987; Odum 1977).

Simplistic theories of environmental determinism based upon models of pre-dictability have given way over the last half century to a clearer appreciation of the reciprocal interactions between environmental change and human societies (Chisholm 1982; Ulanowicz 1986; White 1988). Scholarship relevant to the interre-lationships between human and environmental systems can be found in geogra-phy, anthropology, engineering, history, political science, sociology, resource and development economics, philosophy, and environmental education. Still needed, however, is a paradigm that directly addresses the interrelationships between the two systems.

The second conceptual area concerns the utilization of knowledge and the planning that should follow. It is a description of the way in which knowledge is val-ued in the context of public choice and participation. C. P. Snow examined why pol-icy makers feel they do not have access to the best available knowledge, while sci-entists believe that policy makers reject the options generated by the best available information in favor of what is politically expedient. Studying science-based policy making, Caplan (1979) concludes that Snow was right; there are two distinct cultures, policy and science, which are readily discernible at the microlevel but obscured at the macrolevel. He argues that different "bridging" strategies are needed to integrate these two distinct levels of making decisions. It is here that anthropologists can contribute by devising bridging strategies appropriate to the different cultures.

But solving our global environmental problems cannot be done by scientists and high-level policy makers alone. Our track record speaks for itself, as illustrated by the multinational lending institutions' project failure rate of around 33 percent. If we do not get the public involved, decisions will be short-lived or aborted at the outset. Debates surrounding the human dimensions of sustainable environmental policy have conceded the need to create knowledge based upon an understanding of human systems. Environment and development programs are now emphasizing the ideology of participation while attempting to expand institutional and individual commitment to project success. We are uneasy with our newfound awareness, knowing we need to make major course corrections, yet we wonder if a more fully informed public would act out of enlightened self-interest to make environmentally sound decisions.

Wallerstein (1984) contends that the dilemmas of collective versus individual interests, or short-term versus long-term interests, are contradictions inherent in the models of public choice and political economy. David Mathews (1984), however, offers a competing theoretical perspective and challenges conventional wisdom by asking "When do we know that the public has made a sound decision?" He asserts that it is "when the public understands what the consequences are and then decides what is to be done." Mathews argues that although the political arena in the United States treats political choice as if it is a matter of practice, in reality, "choice is a matter of valuation." The overview on environmental change in the introduction of this article touches upon all of these perspectives closely corresponding to Mathews's paradigm of broad public responsibility in the decision-making process.

Sustaining the Dance of Development

Unfortunately, policy makers usually regard economically oriented development and the "environmentally oriented concept of sustainability" as "two extremes" in the range of policy options. According to Nijkamp and Soeteman (1988), development is largely operationalized in relation to economic theory. However, they argue that an approach oriented to broader goals and shared common interests assists us in moving beyond a narrow environment-development operational framework. On the other hand, Daly and Cobb (1989) argue that more attention must be paid to the natural sciences and that the first two laws of thermodynamics must be the starting point for creating all economic structures. Each of these models suggests a redefinition of the field of economics. Sustainable development in such revised economic terms will better reflect the question of how can we have development and be responsive to ecological systems.

The great question, then, is to what degree homo sapiens can use the environment without degrading or permanently altering it. Agricultural systems that are directly dependent upon essential biological functions and environmental resources illustrate very clearly how preservation and conservation affects sustainability and food availability (Brush 1986; Odum 1984). In studying tropical rainforests and global ecosystems, Raven (1990) reminds us how poorly we understand complex biological systems, for example, the rates and causes of species extinction. He is also quick to point out that genetic engineering can transform but not create genetic material.

Sustainability is a difficult concept to comprehend in precise terms. Consequently, within the environmental community there is disagreement about whether natural factors or human action should be assigned priority in sustaining and maintaining systems. There is agreement that a harmonious balance is needed that will influence the survival and maintain the dynamics of natural systems. At what level and how are other matters.

For example, Western (1984) emphasizes the necessity to change human perceptions and strengthen the receptiveness of individuals to "new methods" and to the value of sustainable practices. Holling (1986), in contrast, concentrates on the need to pay more attention to the natural capacities of ecosystems for resilience and recovery as an impetus for creating broad sustainable patterns for development. In working on the global biodiversity crisis, Miller (1984) argues that the way in which policies are decided and implemented are a barrier to achieving sustainable environmental goals. He suggests it is the juncture where philosophy and knowledge combine to define alternatives that demarcates effective decisions to achieve desired goals. Taking a more mediating approach, Barbier (1987) suggests that sustainability is an ongoing process of dynamic and adaptive trade-offs that maximize goals across human and environmental systems. Grumbine (1992) challenges this ongoing discussion of adequate and inadequate, appropriate and inappropriate approaches by framing a three-pronged strategy proposing (1) that all levels of policy formation incorporate conservation biology as a standard frame of reference; (2) that decision making about land use be based on broad participation at the local and regional levels; and (3) that a biocentric ethic be developed to guide personal and institutional behavior.

In the broader view of the international environmental arena, major policy efforts have attempted to embrace the notion of dynamic interrelationships. These efforts strive to link competing interest groups who can address pressing environmental issues such as damage to the ozone layer, acid rain, the energy crisis, water quality, and biological diversity. The 1980s saw a proliferation of local enterprises where grassroots organizations pressured national governments to place the environment on their national development and foreign policy agendas (Brown 1990). Citizens impatient with government apathy continue to act on their own behalf as seen in the proliferation of nongovernmental organizations (NGOs) and the involvement of NGOs in planning and implementing environmental policy.

In the early 1990s the Union of Concerned Scientists made public a four-page document signed by more than 100 Nobel laureates warning that the human impact on the planet is leading to irreversible damage of the natural and human environment. This heightened public awareness has resulted in governmental agencies' becoming serious about their environmental policy efforts (Clark and Munn 1986; Given 1988; Goodland 1985; Munn 1987; Redclift 1987; World Commission on Environment and Development 1987). Even those governments that have been antagonistic toward any discussion of the complexity of environmental degradation and destruction are now beginning to engage in purposeful discussions around global ecosystems issues as illustrated by the Tropical Forestry Action Plan (World Resources Institute 1987), the 1992 United Nations Conference on Environment and Development (UNCED) in Rio de Janeiro, Brazil, and, post-UNCED, the formation of the Northeast Asia Environmental Cooperation Network.

Development and natural resource management strategies are being shaped from within different frames of knowledge, at all levels—the local people who utilize but often do not control resources, governmental officials who oversee the resources, and the top decision makers and policy makers who address questions on natural resource issues (Brokensha 1987; Burch 1987, 1988). Taking this one step further, policy and planning options are now being defined in terms of programs supplemented by projects that are driven by both top-down and bottom-up priorities. Each step in this process requires approaches that encourage public choice and broad public participation consistent with institutional and social values. It is, therefore, imperative to know how to use and communicate anthropological insights effectively in the process of solving environmental problems (Bunting and Wright 1984; Jacobs and Monroe 1987; Puntenney 1983, 1990; Puntenney and Stapp 1981; Rappaport 1994; Widstrand 1976).

The authors in this volume, though eclectic in their approaches to environmental theory, all see their role as giving voice to seriously compromised global ecosystems and impacted human populations. It is important to note that the weight of the argument presented is that behavioral solutions and technological fixes are not enough. There must also be an understanding of the dynamic interdependencies.

Knowledge Utilization and Planning

Policy makers are increasingly confronted with complex and threatening syndromes that often extend beyond local and regional to national boundaries. A policy that effectively anticipates and addresses pervasive environmental issues is one of "Ready, Aim, Fire," instead of the usual crisis management approach of "Ready, Fire, Aim" (Mac Neil 1988). Decision makers want to have access to environmental knowledge in its various forms—technical, scientific, economic, social/cultural, political, ecological—and transform this information, both internally and externally, into the wider policy-making arena. Unfortunately, most studies of interactions between development and environment have focused primarily on isolated problems such as soil erosion, deforestation, acid deposition, water quality, and desertification, characterizing, more or less, the immediate impact and ameliorative measures (Clark 1988). Valuable though these are, they do not provide decision makers with the tools to formulate broader plans. For both the decision maker and the practitioner, studies of particular causes and effects offer only limited information on which to formulate and implement strategic plans that anticipate and address the interdependencies of ecological and human systems.

Policy makers not only suffer from lack of good information, they are also burdened by a traditional fragmentary approach to public policy. Focusing on investment-oriented approaches to economic development or technical solutions or expedient political priorities cannot delineate appropriate options or provide useful insights on how to determine a sustainable level of use (Forester 1988; Hart 1983; Michael 1973; Schon 1980; Wallerstein 1984). The consequences of administrating from a narrow perspective applied to poor information has been seen, for example, in the failure of investment projects to develop the Brazilian rainforests (Davis 1977; Hecht and Cockburn 1990; Schmink and Wood 1987).

Successful policy making prescribes an integrated approach of agenda setting and alternatives for public policy (Kingdon 1984). Therefore, within the present context of global ecosystems, it is essential to understand how to incorporate perspectives and attitudes from entire policy communities drawing upon diverse expertise from NGOs, governmental organizations, private and nonprofit groups, think tanks, and the academic community to understand the meaning of public policy. Correspondingly, insights are badly needed that re-examine the meaning of public involvment in diagnosing and mediating environmental problems.

Engaging a Dialogue: A Role for Applied Anthropology

We know we have developed a fair amount of scientific and technical knowledge. On another level, we have made real progress in sorting out the application of practical knowledge in such areas as environmental management, community-based development and improved production systems (DeWalt 1988; Orlove 1980). It is between these levels, where managerial and scientific knowledge meet in the context of political and social systems, that things are murky.

It is here in this zone of ambiguity and ambivalence that the application of anthropological perspectives might make its most significant contribution to public policy making. The discipline offers an approach that can identify "condition, context, and process" (Eddy and Partridge 1987; Nader 1988; Rappaport 1979; Spicer 1976; Trotter 1990; van Willigen 1984).

Anthropologists Sherry Ortner and Roy Rappaport provide an essential intellectual foundation from which professionals can build upon in trying to make sense of human-environmental interrelationships. Ortner (1984) denotes an important shift toward understanding society and culture as constructed by human experience. She argues that "a new key symbol of theoretical orientation is emerging, which may be labeled 'practice' (or 'action' or 'praxis')."

In the past, disengaged notions of "Society is a human product. Society is an objective reality. Man is a social product" (Berger and Luckman 1967:61) dominated the study of human systems, with anthropology primarily emphasizing the second construct. Ortner observes, however, that the modern versions of practice theory accepted "all three sides of the Berger and Luckmann triangle, that society is a system, that the system is powerfully constraining, and yet that the system can be made and unmade through human action and interaction," reflecting a tendency across the human sciences toward the study of complex systems through praxis. The significance of these trends would not be fully understood until the 1990s, after developments in computer technology and satellite imagery would allow us to model complex systems and the paradigm shift in ecology that would alter our view of society further.

What we have learned about the human impact on global environmental issues and related dynamic interactions with the ecosphere suggests a more encompassing operational model that embodies sociocultural paradigms. What we have not understood well is where and how in the late 20th century humans fit in that larger global perspective. Rappaport (1994) articulates this state of affairs, charging anthropology with the task of making "discourses intelligible and audible" and stating that anthropology "is best able of the social sciences to 'give voice' to 'Others' or, better, to help others translate their understandings into terms that people other than they themselves can understand."

The anthropologists who are engaging in the current dialogue among environmental professionals, the development advocates, and other interested parties (from local institutions to NGOs and from local leaders to government officials to international authorities) are beginning to grasp the dynamics of the problems and to define the core components of the solutions that will allow us to alleviate the impact of humans on the environment.

The professionals who have been selected to contribute to this volume bring different institutional perspectives on how best to make use of anthropological knowledge within the environmental community. They have learned first hand about applying a process-based model to environmental issues. Each represents the voices of tomorrow in the field of environmental anthropology. Michael Wright discusses wildlife conservation and human development in Africa, arguing the merits of a flexible model for shaping environmental policy and practice. Michael Painter stresses the need for anthropologists to incorporate the concept of environmental degradation both in the construction of their theoretical models and in the application of these models to practice. He argues that research and training must be placed "in the context of the political debate" that defines alternatives and the capacity to act. David Gow defines the role for the practicing anthropologist in confronting environmental issues in the development arena. Like Painter, Gow strongly urges anthropology to acknowledge the primacy of the political arena for solving international environmental problems. He believes that anthropologists must communicate their insights not only to decision makers but also to fellow anthropologists. Robert Winterbottom analyzes the Tropical Forestry Action Plan (TFAP), initiated by the World Resources Institute. Through the TFAP, international agencies hope to create a new kind of institutional structure to bring nations together to reverse tropical deforestation. The original intent was to formulate a mutually beneficial policy, but the process of developing an appropriate model has stalled under strong criticism, and the plan has been restructured with recommendations from Winterbottom to incorporate critical anthropological perspectives. Alberto Costa et al. examine environmental awareness in Brazil describing the linkages between culture and ethno-ecology and the emergence of environmental awareness and risk perception. They analyze the difficult theoretical, political, and ethical problems they had to negotiate in carrying out their work. Finally, Terence Turner's commentary on each of the articles provides an important critical framework from which the discipline of anthropology can better understand the world of environment affairs. Drawing upon his own work in Brazil with the Kayapo Indians, he pinpoints the sources of structural contradictions and conflicts of interest within environment and development priorities while evaluating the important role of the discipline. He concludes that anthropologists must consider the future direction of anthropology in terms of environment and development.

It is hoped that anthropologists and colleagues within the environmental community will find useful insights from these experts to interpret and guide programs, policy development, and institutional priorities as applied to environmental-human interrelationships.

References Cited

Akiyama, Toshiko
    1993 Development of Japanese Environmental Policy and its Implications for Asian Countries. Paper presented at the international workshop on Development and the Environment: The Experiences of Japan and Industrializing Asia. Tokyo: Institute of Developing Economies.
Arizpe, Lourdes
    1991 The Global Cube. International Social Science Journal 43:600–608.

Barbier, Edward B.
    1987 The Concept of Sustainable Economic Development. Environmental Conservation. Lusanne, Switzerland: Elsevier Sequoia.
Berger, Peter, and Luckmann, Thomas
    1967 The Social Construction of Reality. Garden City, NY: Doubleday.
Brokensha, David W.
    1987 Development Anthropology and Natural Resource Management. L'Uomo 11(2):225–250.
Brown, Janet
    1990 Global Briefing V: The Environment/Development Connection or the Critical Importance of the Developing countries for the Sustainability of Global Systems. Washington, DC: World Resources Institute.
Brush, Stephen B.
    1986 Genetic Diversity and Conservation in Traditional Farming Systems. Journal of Ethnobiology 6(1):151–167.
Bunting, Bruce, and R. Michael Wright
    1984 Annapurna National Park, the Nepal Plan for Joining Human Values and Conservation of a Mountain Ecosystem. Washington, DC: World Wildlife Fund.
Burch, William R., Jr.
    1987 Gods of the Forest—Myth, Ritual and Television in Community Forestry. Paper presented at the Regional Community Forestry Training Center, Asia-Pacific Seminar, Bangkok, Thailand.
    1988 The Uses of Social Science in the Training of Professional Social Foresters. Journal of World Forest Resource Management 3(2):73–109.
Caplan, Nathan
    1979 The Two-Communities Theory and Knowledge Utilization. American Behavioral Scientist 22(3):459–470.
Chisholm, Michael
    1982 Modern World Development. London: Hutchinson.
Clark, William C.
    1988 The Human Dimensions of Global Environmental Change. A report prepared for the U.S. National Research Council's Committee on Global Change as a Contribution to the Preliminary Plan for U.S. Participation in the International Geosphere-Biosphere Program. Unpublished draft, Harvard University, Kennedy School of Economics.
Clark, William C., and Robert E. Munn, eds.
    1986 Sustainable Development of the Biosphere. Cambridge: Cambridge University Press.
Cole, Lamont C.
    1958 The Ecosphere. Scientific American 198:83–92.
Colson, Elizabeth
    1985 Using Anthropology in a World on the Move. Human Organization. 44:191–196.
Daly, Herman E.
    1990 Toward Some Operational Principles of Sustainable Development. Ecological Economics 2:1–6.
Daly, Herman E., and John B. Cobb Jr.
    1989 For the Common Good: A Sustainable Future. Boston: Beacon Press.
Davis, Shelton H.
    1977 Victims of the Miracle: Development and the Indians of Brazil. Cambridge: Cambridge University Press.
DeWalt, Billie R.
    1988 Halfway There: Social Science in Agricultural Development and the Social Science of Agricultural Development. Human Organization 47(4):343–353.
Eddy, Elizabeth M., and William L. Partridge, eds.
    1987 Applied Anthropology in America. New York: Columbia University Press.
Engel, J. Ronald, and Joan Gibb Engel
    1990 Ethics of Environment and Development: Global Challenge, International Response. Tucson: University of Arizona Press.
Escobar, Arturo
    1991 Anthropology and the Development Encounter: The Making and Marketing of Development Anthropology. American Ethnologist 18:658–682.

Forester, John
    1988 Beyond the Equity-Efficiency Conflict: The Practical Analysis of Ambiguity in Plan-
    ning Practice. The Journal of Architectural and Planning Research 5(2):91–109.
Gallopin, Gilberto C.
    1991 Human Dimensions of Global Change: Linking the Global and the Local Processes.
    International Social Science Journal 43:707–718.
Given, David R.
    1988 Conserving Biological Diversity on a Global Scale. Paper presented at the 35th
    Annual Systematics Symposium "Conserving Biological Diversity: Prospects for the 21st
    Century."
Gomez-Pompa, Arturo, and Andrea Kaus
    1990 Taming the Wilderness Myth: A View of Environmental Education from the Field. La
    Educacion Ambiental en Mexico en la Decada de los Noventas. Paper presented at the
    American Association for Environmental Education Annual Meeting, San Antonio, Texas.
Goodland, Robert
    1985 Wildland Management in Economic Development. Washington, DC: The World
    Bank.
Gow, David D.
    1991 Collaboration in Development Consulting: Stooges, Hired Guns, or Musketeers?
    Human Organization 50(1):1–15.
Grumbine, R. Edward
    1992 Ghost Bears: Exploring the Biodiversity Crisis. Washington, DC: Island Press.
Hackenberg, Robert
    1976 Scientists or Survivors?: The Future of Applied Anthropology under Maximum
    Uncertainty. In Do Applied Anthropologists Apply Anthropology? Michael V. Angrosino,
    ed. Pp. 118-133. Southern Anthropological Society Proceedings No. 10. Athens: Univer-
    sity of Georgia Press.
Hart, Stuart L.
    1983 Strategic Problem Solving in Turbulent Environments: A Description and Evaluation.
    Ann Arbor: University of Michigan.
Hecht, Susanna, and Alexander Cockburn
    1990 The Fate of the Forest: Developers, Destroyers and Defenders of the Amazon. New
    York: Harper Perennial.
Holling, C. S.
    1986 Resilience of Terrestrial Ecosystems: Local Surprise and Global Change. In Sus-
    tainable Development of the Biosphere. W. C. Clark and R. E. Munn, eds. Pp. 292–320.
    Cambridge: Cambridge University Press.
Jacobs, P., and D. Monroe, eds.
    1987 Conservation with Equity: Strategies for Sustainable Development. Ottowa: Canada
    Department of the Environment.
Johnson, Jeffrey C., and Michael K. Orbach
    1990 A Fishery in Transition: The Impact of Urbanization on Florida's Spiny Lobster
    Fishery. City & Society 4(1):88–104.
Kingdon, John W.
    1984 Agendas, Alternatives, and Public Policies. Glenville, IL: Scott, Foresman and
    Company.
Kottak, Conrad P.
    1990 Culture and "Economic Development." American Anthropologist 92:723–731.
Leach, Edmund
    1968 A Runaway World? New York: Oxford University Press.
Lovelock, James E.
    1990 Hands up for the Gaia Hypothesis. Nature 344:100–102.
McCay, Bonnie J., and James M. Acheson, eds.
    1987 The Question of the Commons: The Culture and Ecology of Communal Resources.
    Tucson: University of Arizona Press.
Mac Neil, J.
    1988 Sustainable Growth. Policy Options Politiques (March):22–25.
Mathews, David
    1984 The Public in Practice and Theory. Public Administration Review 44 (Special
    Issue):120–125.

Mathews, Jessica Tuchman
   1991 The Implications of U.S. Policy. *In* Preserving the Global Environment: The Chal-
      lenge of Shared Leadership. Pp. 309–323. New York: Norton and Company.
Michael, Donald
   1973 On Learning to Plan and Planning to Learn. San Francisco: Jossey-Bass.
Miller, Kenton
   1984 National and Regional Conservation Strategies. *In* Sustaining Tomorrow: A Strategy
      for World Conservation and Development. Francis R. Thibodeau and Herman H. Fields,
      eds. Pp. 69–76. Hanover and London: University of New England.
Munn, Robert E.
   1987 Environmental Prospects for the Next Century: Implications for Long-Term Policy
      and Research Strategies. Laxenburg, Austria: International Institute for Applied Systems
      Analysis.
Nader, Laura
   1988 Post-Interpretive Anthropology. Paper presented at the American Anthropological
      Association Annual Meeting, Phoenix, AZ.
Nijkamp, Peter and Frits Soeteman
   1988 Ecological Sustainable Economic Development: Key Issues for Strategic Environ-
      mental Management. International Journal of Social Economics 15:3–4.
Odum, Eugene P.
   1977 The Emergence of Ecology as a New Integrative Discipline. Science 195:1289–
      1293.
   1984 Properties of Agroecosystems. *In* Agricultural Ecosystems: Unifying Concepts.
      Richard Laurence, Benjamin Stinner, and Garfield House, eds. Pp. 5–11. New York: John
      Wiley and Sons.
Orlove, Benjamin S.
   1980 Ecological Anthropology. Annual Review of Anthropology 9:235–273.
Ortner, Sherry B.
   1984 Theory in Anthropology since the Sixties. Society for Comparative Study of Society
      and History 26:126–166.
Puntenney, P. J.
   1983 The Unity of Learning: A Systems Perspective on Environmental Education. Ann
      Arbor: University of Michigan.
   1990 Defining Solutions: The Annapurna Experience. Cultural Survival Quarterly 14(2):9–
      14.
   1994 Global/Local Environmental Agendas: Problem Solving in Northeast Asia. Paper
      presented at annual meeting of the Society for Applied Anthropology, Cancun, Mexico.
Puntenney, P. J., and William B. Stapp
   1981 National Strategy for Environmental Education: A Planning and Management Proc-
      ess. Environmental Education and Information 1(1):39–48.
Rappaport, Roy A.
   1979 Ecology, Meaning, and Religion. Richmond, CA: North Atlantic Books.
   1994 Disorders of Our Own: A Conclusion. *In* Diagnosing America: Anthropology and
      Public Engagement. Shepard Foreman, ed. Pp. 235–294. Ann Arbor: University of
      Michigan Press.
Raven, Peter
   1990 Comments on the Brazilian Rainforests and Global Ecosystems. Paper presented
      at the World Bank Forum on Conservation and International Development, Washington,
      DC.
Redclift, Michael
   1987 Sustainable Development: Exploring the Contradictions. London: Methuen.
Rowe, J. Stan
   1961 The Level-of-Integration Concept and Ecology. Ecology 42:420–427.
Rowe, J. Stan
   1989 What on Earth is Environment? The Trumpeter 6(4):123–126.
Schmink, Marianne, and Charles H. Wood
   1987 The Political Ecology of Amazonia. *In* Lands at Risk in the Third World. P. D. Little,
      M. M. Horowitz, and A. E. Nyerges, eds. Pp. 38–57. Boulder, CO: Westview Press.
Schon, Donald
   1980 The Reflective Practitioner. Cambridge, MA: MIT.

Schnneider, S. H.
1989   Global Warming. San Francisco, CA: Sierra Cub Books.
Soule, Michael E., ed.
1986   Conservation Biology: The Science of Scarcity and Diversity. Sunderland, MA: Sinauer.
Spicer, Edward H.
1976   Anthropology and the Policy Process. In Do Applied Anthropologists Apply Anthropology? Michael V. Angrosino, ed. Pp. 118–133. Southern Anthropological Society Proceedings No. 10. Athens: University of Georgia Press.
Sponsel, Leslie E.
1986   Amazon Ecology and Adaptation. Annual Review of Anthropology 15:67–97.
Stapp, William B.
1984   Building Support for Environmental Education. In Sustaining Tomorrow: A Strategy for World Conservation and Development. Francis R. Thibodeau and Hermann H. Field, eds. Pp. 87–93. Hanover and London: University of New England.
1986   An International Definition of Environmental Education. Columbus, OH: ERIC.
Tansley, Arthur G.
1935   The Use and Abuse of Vegetational Concepts and Terms. Ecology 16:284–307.
Trotter, Robert T.
1990   The Challenge of Anthropological Practice: Trends, Ideals, Speculations, and the Joys of Getting Things Done. Keynote presentation for the 10th Annual Meeting of the High Plains Society for Applied Anthropology, Northglenn, CO.
Ulanowicz, Robert E.
1986   Growth and Development: Ecosystems Phenomenology. New York: Springer-Verlag.
van Willigen, John
1984   Truth and Effectiveness: An Essay on the Relationships between Information, Policy and Action in Applied Anthropology. Human Organization 43:277–282.
Wallerstein, Immanuel
1984   The Development of the Concept of Development. In Sociological Theory. Randall Collins ed. Pp. 102–116. San Francisco: Jossey-Bass.
Western, David
1984   Conservation Based Rural Development. In Sustaining Tomorrow: A Strategy for World Conservation and Development. Francis R. Thibodeau and Hermann Field, eds. Pp. 94–108. Hanover and London: University of New England.
White, Gilbert F.
1988   Greenhouse Gases, Nile Snails, and Human Choice. Distinguished Lecture Series on Behavioral Science, Institute of Behavioral Science. Boulder: University of Colorado.
Widstrand, C.
1976   Rural Participation and Planning. In Development from Below. D. Pitt, ed. Pp. 139–144. Hague: Mouton Publishers.
Wilson, E. O.
1988   Biodiversity. Washington, DC: National Academy Press.
World Commission on Environment and Development
1987   Our Common Future. New York: Oxford University Press.
World Resources Institute
1987   Tropical Forestry Action Plan. Washington, DC: FAO/WRI/World Bank/United Nations Environment Program.
Wright, R. Michael
1993   The View from Airlie. Summary of the Community Based Conservation Workshop held in Airlie, VA. Liz Clairborne and Art Ortenberg Foundation.
Wulff, Robert, and Shirley Fiske, eds.
1987   Anthropological Praxis: Translating Knowledge into Action. Boulder, CO: Westview Press.

Alleviating Poverty and Conserving Wildlife in Africa: An "Imperfect" Model from Zambia

R. Michael Wright

The Perfect Model That Never Was

> Whenever the British have undertaken the development of a new country, amongst the earliest regulations will nearly always be found a measure which is designed to afford a degree of protection to the local fauna. It is a pleasing British characteristic. Accordingly, since the last years of the 19th century the well-being of much of East Africa's splendid wildlife was carefully fostered by regulations, which not only puzzled the indigenous population, but was an irritant. [Speech of Captain Charles R. S. Pitman before the Mammal Society of the British Isles, March 27, 1960, quoted in Marks 1984]

Nature's eternity is symbolized in the pristine African plains teeming with spectacular herds of wildlife: wildebeest from horizon to horizon, majestic lions roaming among stately giraffes, scampering warthogs, their tails erect, and hyenas skulking through the underbrush. In the popular image, the African landscape is generally devoid of the African people.

> The problem is rooted in the nature of the colonial relationship itself, which allowed Europeans to impose their image of Africa upon the reality of the African landscape. Much of the emotional, as distinct from economic, investment which Europe made in Africa has manifested itself in a wish to protect the natural environment as a special kind of "Eden," for the purpose of the European psyche, rather than as a complex and changing environment in which people have actually had to live . . . the mythology of the African environment and the symbol of Africa as a yet unspoiled Eden continues to stimulate many of those who wish to intervene in the way the environment managed in Africa. [Anderson and Grove 1987]

This early history would be only of passing interest were it not for the astonishing continuity in policy from the colonial state to the independent governments of modern Africa. The myth is maintained even today through a deluge of beautiful coffee table nature books on Africa and nightly nature specials with widespread and continuing popularity. Similarly, the continuing enthusiasm for outside intervention is seen most recently and dramatically in the calls from the capitals of Europe and the United States for a ban on all trade in elephant ivory, even for those countries such as Zimbabwe and Botswana, which are acknowledged to have successfully managed their populations.

Some have argued that the "Eden" which European mythology sought to preserve never existed at all. Colonists arriving in East Africa in the 1890s found what appeared to be a vast, empty paradise, unaware that it may have been made so by smallpox, which devastated the human population, and rinderpest, which had annihilated the cattle herds (Deihl 1988). Similar patterns occurred elsewhere on the continent: in the Belgian Congo, for example, a population of 40 million in 1880 declined to 9.25 million by 1933 (Bell 1987). In this apparent void, the "hunting-mad, animal-loving British" and, more recently, African national governments have set aside areas for protection of wildlife and natural systems. In Tanzania, Botswana, Zambia, Zimbabwe, and Senegal, even in Rwanda, one of the most densely populated countries in Africa, these systems exceed 10 percent of the gross land area. Whereas the first U.S. national parks were established to preserve monumental landscapes, the protected areas of Africa focused from the beginning on wildlife, initially preserving the royal hunting prerogative from colonizer's excesses and, after World War II, turning toward nature preservation. Although the motives for their creation evolved, these protected areas were generally carved out of communal, tribal lands whose populations were treated, more often than not, as irrelevant.

Just as the land in parks was alienated from the communities and maintained solely for the use of the government, so, too, did the colonial regimes reserve wildlife for the use of the ruling white authority. With independence, the racial makeup of the beneficiaries changed, but the animals remained the property of the state, and revenues generated from safari and other forms of utilization went to the national treasury. The rural subsistence farmers who coexisted with wildlife developed a pattern of uncontrolled illegal usage. Wildlife became a classic case of an open-access resource. Local people perceived little likelihood of any future benefit from deferred utilization. On the other hand, the government, claiming exclusive ownership, lacked the political, financial, and human capacity to enforce their rights and to protect the resource.

The Price of Perfection

[The Ik] too were driven by the need to survive against seemingly invincible odds, and they succeeded, at the cost of their humanity. . . . The Ik have relinquished all luxury in the name of individual survival, and the result is that they live on as a people without life, without passion, beyond humanity. [Turnbull 1972]

The Ik, banished from their traditional hunting areas and forced to scratch out something resembling a life on barren hillsides, are just one of the better-documented examples of the social and cultural disintegration of a people expelled to make way for a protected area. Similar cases exist with the San Bushmen (Volkman 1986), the Maasai (Parkipuny and Berger 1993), and also outside the African continent (Clay 1985; Gardner and Nelson 1981; Lawson 1985; Poole 1989). As William Partridge (1991) demonstrates in his analysis of involuntary resettlement, both proponents of development and conservationists have largely ignored the social consequences of their actions. Conservation decisions, too often taken as though there were no people in Africa, have deprived rural peasants of land, traditional food sources, and the ability to trade in wildlife products, and have severed key cultural links that hunting played in many societies. Rural people have been forced to compete with wildlife for grazing and water and to endure destruction of their crops and loss of human life. The simple fact is that except for the financial costs of administration, the burden of Africa's extensive protected area system has fallen on the poor rural population while the inspirational, educational, recreational, and scientific benefits have accrued primarily to foreigners and the prestige and revenues to national governments (Bell 1987).

The human cost of the alienation of land in protected areas has escalated as the population of sub-Saharan Africa has grown—an increase of 80 million people between 1970 and 1985. A shrinking land base for subsistence agriculture, overgrazing, and a shortage of fuelwood for energy have created deforestation and soil erosion. All this has been happening in a region whose carrying capacity has already been exceeded in many of the more arid countries. Deterioration on their periphery affects the integrity of the protected areas themselves, as most suffer from design flaws such as boundaries that cut across ecosystems, over migration routes, and through traditional wildlife grazing areas.

As nearby peoples' own resource bases deteriorate, the parks are ever more attractive. Exclusion through punitive measures becomes ever more financially and politically costly. Here lies the seeds of the destruction of Africa's "Eden" and the link that makes the villager and poacher allies. The future of protected areas and the conservation of wildlife outside the borders of parks, and indeed perhaps within them as well, depend on solving two problems: first, the question of equity—how to provide enough resources to allow surrounding farmers to achieve a reasonable standard of living; second, whether institutions to protect wildlife can be found or created that tap enlightened self-interest of African people themselves. Only conservation strategies grounded in the traditional structure of rural Africans will be enforced by them rather than through a central policing system that few governments in Africa today can afford. The latter question defines the role of and challenge to the anthropological profession in the conservation movement.

An "Imperfect" Alternative

Better to let them do it imperfectly than to do it perfectly yourself, for it is their country, their way, and your time is short.

—Lawrence of Arabia

Culturally based ecosystem management systems (Matowanyika 1989) must be developed that link conservation to a process of rural development and to the survival of agrarian societies in Africa. Needed are alternatives that look at the wildlife resources flowing out of and surrounding parks as a basis for community management systems. Only a culturally based conservation system that meets both ecological and social criteria can succeed in Africa. Conventional development has often ignored the ecological criteria, and, until recently, conservationists have ignored the social and cultural parameters.

In conventional development thinking wildlife is considered a less reliable resource than domesticated animals. Wild species, however, have definite ecological advantages over traditional livestock. In semiarid savannas they are more selective and have complementary grazing preferences, are more tolerant of droughts, and are less likely to degrade the productive potential of ecosystems than cattle. These same multiple values can be realized at lower stocking levels, thus posing less risk to ecosystem stability. As we are now learning, wildlife have a comparative economic advantage that has been distorted by a subsidized pricing and marketing system heavily prejudiced in favor of cattle (Child and Nduku 1985).

Systems using wildlife have become well established in private ranches in Zimbabwe (Cumming 1987) and, to a lesser extent, Kenya (Hopcraft 1986). Commercial ranchers have found that harvesting wildlife can be more profitable than cattle under semiarid conditions on marginal land, an advantage made even greater when foreign exchange factors are considered. As a result, a majority of commercial cattle ranchers in Zimbabwe now devote at least some of their lands to wildlife, which often provides 25–50 percent of net income. Contrary to the enthusiastic claims of some wildlife ranching proponents, the Zimbabwe experience indicates that wildlife is not generally a cheaper source of protein than domestic stock (although in some instances it can produce comparable meat with less risk of eco-system degradation). The real economic attraction of wildlife is that wild popula-tions can sustain an array of simultaneous income generating uses—culling or cropping for meat production, preparation or marketing of wildlife products, safari hunting, and nonconsumptive uses such as game viewing and photographic safa-ris. Such diversification without intensification of resource use is the essence of sus-tainable development (Development Alternatives 1988; Muir 1988).

Can the experience of the private lands be translated to communal proprietor-ship in areas adjacent to and surrounding national parks and protected areas? For millennia, wildlife hunting has been a major component in the diets of rural commu-nities in Africa and elsewhere in the developing world (Caldecott 1986; Clay 1988; International Union for the Conservation of Nature 1981; Redford and Robinson 1987). The use of wildlife for protein in western and central Africa has been particu-larly well documented (Asibey 1974; Hart and Petrides 1987; Ntiamoa-Baidu 1987). Tribal social structure and culture in southern Africa also include patterns of allocat-ing and managing wild species, which were disrupted during the colonial period (Cumming and Taylor 1989; Hitchcock 1989; Murphree and Murombedzi 1987; Owen-Smith and Jacobsohn 1989; Selpapitso 1988). In many parts of rural Africa these traditional authority structures, though seriously weakened in the early inde-pendence period, are still in place and could be revived by regimes of local proprie-torship of wildlife. From the conservation point of view, giving local communities a stake in wildlife could do much to discourage poaching and reverse the currently adversarial relationship between government ministries responsible for wildlife pro-tection and their poor human neighbors.

Practicing the "Imperfect"

This approach of community participation in conservation confirms what we already know, namely, that wildlife resources are wealth. But the significance of the programme lies in the fact that it will ensure that the wealth realized is used for the benefit of the people to whom the resources belong instead of satisfying the individual and selfish needs of an indiscriminate poacher.

This is how, in July 1989, Dr. Kenneth D. Kaunda, then-president of the Repub-lic of Zambia, described the Administrative Management Design for Game Man-agement Areas (ADMADE), one of the growing number of experiments in Africa seeking a balanced approach to wildlife protection. ADMADE is a national program built upon the experience of Lupande Development Pilot Project (LDP). ADMADE has been developed by the Zambia National Parks and Wildlife Service over the past eight years.

The Setting

Zambia contains some of the largest remaining concentrations of African sa-vanna wildlife, and isolated rural populations depend heavily on these resources. Thus, Zambia has the potential to design a multiple use conservation approach that is an inseparable part of its traditional culture. Zambia has a system of 19 national parks covering 8 percent of the country, which the government seeks to manage according to strict protection standards. Buffering these parks are 32 game man-agement areas (GMAs) covering an additional 22 percent of the country. GMAs were established for multiple use and local human benefits; however, without a government extension program, the GMAs have not yet contributed significantly to local economies.

An exception is in the Luangwa Valley, in eastern Zambia, where the LDP is located. The valley contains about 25,000 elephants, a remnant population of black rhinos, as well as large numbers of buffalos, hippos, crocodiles, and a full range of predators and small game. The valley has four national parks surrounded by sparsely settled GMAs, which together cover some 80 percent of the catchment area containing significant wildlife populations and supporting an important safari hunting industry. Most of the 40,000 people of Luangwa Valley from five distinct tribes are poor subsistence farmers living in scattered villages along alluvial drainages. Social services, such as schools and health centers, are inadequate; malnutrition and protein deficiencies are common. Little investment has been made in agriculture, and development projects have passed them by. An area that is infested with tsetse fly and where soil conditions make only 3–7 percent of the land arable must accommodate a growing population and its need for food (Marks 1984).

The World Wildlife Fund (WWF), through its Wildlands and Human Needs Program (Wright 1988), supports the expansion of the LDP, which serves as a pilot program and training center, into nine other GMAs.

Evolution of ADMADE

The most important precursor to ADMADE was the Lupande Development Workshop held at Nyamaluma in September 1983. Addressing workshop participants, Chief Malama complained bitterly about the lack of benefits from wildlife on his land:

> We are honest people who are keepers of the wildlife. We do not like poaching and we have been keeping the animals here a long time for the Government, but we receive no benefit for this service. If I beg help for building a clinic or grading our road, the government refuses. Yet, this is the area where both the government and private individuals benefit from wildlife. Tourists come here to enjoy the lodges and to view wildlife. Safari companies come to kill animals and make money. We are forgotten. [Malama 1984:8]

The complaints were well founded as less than 1 percent of safari hunting revenue was returned to local economies (Atkins 1984). Although the complaints were neither new or confined to Zambia, they found a responsive ear in Gilson Kaweche and Akim Mwenya, then the chief research officer and the acting deputy director, respectively, of the National Parks and Wildlife Service (NPWS), and Dale Lewis, who was conducting elephant research with support from the New York Zoological Society. Lewis had begun to listen to local communities after hostile encounters with Chief Malama, who feared that the elephant research was an excuse to expand the adjacent park (Lewis 1989). Mwenya and Kaweche had been smarting under international criticism of NPWS management of Zambia's parks, much of it from former colonial wardens, and the partial dismemberment of the department to establish privately funded antipoaching forces, such as the Save the Rhino Trust. Although these attempts at international policing proved to be equally unsuccessful in stemming the slaughter of elephants, which suffered annual losses of 6–8 percent during the 1970s and early 1980s, and the virtual elimination of rhinos, the time seemed propitious for a new approach. Aware of similar concepts being explored in Zimbabwe through the CAMPFIRE program, the LDP was undertaken as an experiment in grassroots, community-based wildlife utilization and has led to AD-MADE.

The Lupande Development Workshop also resulted in a consultancy that developed the Luangwa Integrated Resource Development Project (LIRDP 1987). A more classic, large-scale integrated project involving sustainable use of agriculture, forestry, fisheries, water, and wildlife, the LIRDP operated in the South Luangwa National Park and the Lupande Game Management area with major funding from NORAD, the Norwegian government's foreign assistance program. Despite their differences in scale, in approach, and in budget, the programs suffer some sibling rivalry as a result of their physical proximity, common roots, and differing relationships with central government departments.

Two governmental decisions in 1983 inadvertently provided the institutional basis for a new approach to wildlife management. Faced with falling copper prices, Zambia's major export and foreign exchange commodity, and a resulting difficulty meeting recurring costs, Zambia government agencies were encouraged to seek alternative funding mechanisms. In response, the Wildlife Conservation Revolving Fund was established authorizing the NPWS to generate and retain revenue and, most important of all, giving the department autonomy over the use of funds generated. The revolving fund provided the critical legal mechanism by which NPWS shares money generated in a GMA with its residents.

Of equal importance, the minister responsible for wildlife designated all classified employees as wildlife officers. The apparently esoteric decision that non–civil servants could be taken on as wildlife officers made possible the development of the Village Scout Program, which employs local villagers and has substantially increased law enforcement, wildlife censusing, and data collection in GMAs. Ordinarily, local villagers would not qualify for government civil service, nor would it be possible for traditional chiefs to be involved in their selection. Village scouts have become a significant, often only, source of employment in rural areas, and, unlike NPWS employees, the scouts can be terminated for nonperformance; they are therefore more highly motivated.

Structure of ADMADE

In sharp contrast to the colonial or traditional conservation approach, AD-MADE postulates that effective conservation and management of wildlife depend upon individuals and communities who share their land with wildlife, becoming full participants in decisions concerning the management and development of the resource and receiving a major share of the revenue derived from exploitation. At the same time, in order for the program to be ecologically sustainable, some of the generated revenues must be set aside in order to support formal wildlife management programs. Without such investments in monitoring and maintenance of the resource base, it will inevitably deteriorate.

The ADMADE structure seeks to balance national level management responsibility for wildlife of the NPWS with systems of local participation at the political, traditional and village levels. The main features of the ADMADE programs are as follows:

- Each GMA and/or hunting block constitutes a wildlife management unit. A civil servant member of NPWS is the unit manager, and staff include NPWS wildlife officers and local village scouts.

- For every wildlife management unit there is a wildlife management author-ity. The authority has the district governor as its chairman. The member-ship includes both political and traditional leadership, including the ward chairmen, chiefs, members of Parliament, and local technical officers. The authority serves as the policy body of the unit. Links to political and tradi-tional leadership provides opportunities to mobilize resources beyond those of NPWS to implement village community projects.
- To ensure that decision making is based on grassroots needs and priori-ties, for every chiefdom within a unit there is a wildlife management sub-authority. Membership includes party branch chairmen, headmen, teach-ers, the unit leader as secretary, clinical officers, and the chief as chairman. The sub-authority determines projects to be funded out of the 35 percent of the revenue available for the village community.
- Revenue generated through safari hunting and any other forms of wildlife utilization is held by the revolving fund and apportioned as follows: (1) 40 percent to the local resource management programs, including employ-ment of local villagers as village scouts; (2) 35 percent to the village com-munity development; (3) 15 percent to operational costs of the NPWS; and (4) 10 percent to ADMADE administration.

Implementation

During 1988 the WWF program focused on support of the LDP as the pilot for ADMADE and its replication in five other GMAs. At the LDP the area patrolled by lo-cal village scouts trained and employed under this program increased by 30 per-cent over the previous year. A 90 percent drop in commercial poaching and a de-crease in snares used for subsistence hunting indicated growing local acceptance. Decreased poaching can be attributed in part to village scout patrols and to local leaders (particularly village headmen) as the program decision body. In 1988 the company that had bid successfully for the safari concession had not yet attained the ADMADE-mandated requirement of 80 percent local employment, but it estab-lished a constructive dialogue with the community and supplied low cost meat. In 1988 a small-scale, locally based culling and tanning program supported by Afri-care employed 20 permanent staff putting K 6,000 per month in salaries into the community (in addition to 40 percent of the annual profits) (Kapungwe and Lewis 1989). Ecological monitoring continued to provide a baseline against which to judge the ecological sustainability of the pilot project.

The ADMADE replication process also began in nine GMAs (five under the WWF Wildlands Program) in 1988. Based on the LDP model, it was estimated that about $50,000 per year would be needed to make a GMA self-sustaining (including capital replacement costs). At that time most GMAs in the ADMADE program aver-aged $30,000 in income exclusively from safari concession fees. The program, therefore, focused on expanding sources of income, such as multispecies wildlife utilization and small-scale tourist enterprise, over the next few years. In addition, as a condition of increased United States Agency for International Development (USAID) funding, the Zambian government agreed that a portion of other safari revenues, such as trophy and license fees, which went to central treasury, would henceforth be shared with the communities where the revenue was earned. Two lo-cal staff biologists were hired in 1988 by NPWS to provide logistical and technical support to unit leaders in the field, and a social scientist was to be sought in 1990.

During this period, 19 unit leaders were trained at Nyamaluma, as were 200 village scout recruits, selected by their chiefs. On July 14, 1989, Zambia President Kenneth Kaunda handed over checks from the revolving fund of K 2.3 million (U.S. $230,000) to nine chiefs representing their share of the revenue derived from game in their areas. This ceremony represented the culmination of the first phase of Zambia's new approach to wildlife conservation.

Discussion

Although the Zambia Wildlands and Human Needs Project is still in the process of evolution, several principles emerge from the experience (Lewis et al., in press).

First, local leadership is irreplaceable. Few projects can succeed without responsive and motivated leaders; in the case of ADMADE, particular reliance has been placed on traditional leadership structure of chiefs and/or headmen and the customs that bind and regulate village communities. Conservation undertaken by central government and imposed by force has proven costly and ineffective in far-flung rural areas. On the other hand, integrating technical and capital inputs from the modern governments with traditional rulers working with the local community has in most chiefdoms proved an effective partnership.

In Latin America, local nongovernmental organizations (NGOs) play a key catalytic role in Wildlands and Human Needs Projects. But there are concerns, particularly in Africa, that internationally supported NGOs and expatriate experts usurp government authority and erode the confidence and morale of both local professionals and village leaders (Ward 1989). Thus, in ADMADE, it is the traditional structure that has played the role that NGOs provide in Latin America. In addition to community leadership, the central government leadership is based upon Zambian professionals rather than swelling the number of foreign experts. Africa already has the largest number of foreign experts per capita in the world, in one estimate 80,000, exceeding the number during the colonial era (Hancock 1989). Nevertheless, the base of Zambian leadership remains exceedingly narrow and represents an institutional point of vulnerability.

No goal of ADMADE is more important than community participation. Only through this mechanism can it be assured that projects are culturally acceptable, sensitive to local needs and aspirations, and responsive to customary laws. The classic approach to conservation, which attempts to exclude human population from strictly protected areas through police action, requires expenditures from U.S. $200–250/Km to be effective (Bell and Clarke 1984); the LDP achieved a 90 percent reduction in elephant poaching at a fraction of that cost ($22/Km) (Lewis et al., in press). If one dare generalize from such a limited sample, wildlife protection costs appear to be inversely related to local participation and benefit. Local village scouts have superior knowledge of areas to be patrolled and lower absenteeism than civil servants. Participation and local employment cause a multiplier effect by taking advantage of Africa's still-prevalent "economy of affection," where the first goal of many successful Africans is to take care of and satisfy the family and the village. The fact that most poachers in the pilot project area come from outside the valley encourages proprietorship of local resources; however, it also raises questions with the replicability of this experience in those areas where poachers are primarily local.

Despite ADMADE's laudable participation, women have not played an equal role. The project has disproportionately benefited men in terms of jobs and bolstered male domination of community decision making. Although women's traditional roles have not included management and utilization of large mammals, women are responsible for providing food, firewood, and medicines and are the primary users of the natural resource base. In addition, women who are in charge of subsistence agriculture suffer more from wildlife depredations in village gardens. This issue still needs to be addressed within ADMADE (Hunter et al., in press).

ADMADE's success ultimately will not depend on meeting the WWF's objectives, or even those of NPWS, but on responding to the felt needs of the communities themselves. Those needs, be they jobs, income, or protein source, must reach a critical threshold at least approximating the opportunity cost of conservation. If the number of beneficiaries is too small, resentment will be bred; as the pilot project demonstrates, however, once a threshold is reached, in theory, peer pressure should take over (Lewis et al., in press).

Modest voluntary contributions from local people, such as labor for the Africare culling program and in-kind or locally raised cash, demonstrate that the projects address perceived local needs and create an increase in the sense of ownership and long-term commitment. Give-away projects or subsidized labor discourage self-reliance and distort the development process. In contrast, village employment in the ADMADE program is directly linked to and paid for by use of the wildlife resource.

A project may have local leadership and community participation, respond to a felt need, and still fail due to an inappropriate legal, institutional, or policy structure. ADMADE is composed of a mixed system including elements of both the common property resource management and government comanagement. Key institutional decisions of ADMADE involve setting up the revolving fund and making it possible to hire non–civil service employees; commitment by the government to return a substantial portion, if not all, revenue from the safari industry to the local community; and sharing control over the wildlife resource. The longer-term question for ADMADE is whether its mixed local-national, private-public approach is able to create true proprietorship over the resource at the community level and escape natural problems of bureaucratic inefficiency and corruption.

A collection of characteristics or elements forms parts of successful grassroots development. The risk to participants from involvement in ADMADE is kept at a minimum, requiring little upfront cash and providing a near-term payoff. Whenever possible, the resources required are already available to the poor, recurring costs are low, and income generation is focused on a reliable, stable market able to support increased production. The safari market appears capable of substantial expansion unless it is undercut by the growth of animal welfare and antihunting sentiment outside Zambia or internal corruption of the licensing system and resource depletion from resident hunting. Capital costs should be kept low. This lesson was demonstrated by the Africare-supported self-help culling project. When an anthrax outbreak interrupted the program there was no danger of external pressure to continue operations in order to pay off loans. Despite this loss of revenue in 1987, total revenue in the pilot project exceeded management costs almost four times, about half of which was used locally, the other half retained by Central Treasury.

Paradoxically, there is a risk in WWF and USAID support of ADMADE. External donor support can damage cultural integrity, local self-confidence, and, thus, motivation. Projects relying on external funds may only progress when such funds are available and may require inflated budgets to meet the support levels required by foreign donors. Ultimately, larger budgets may not be sustainable without depleting the wildlife resource; in the meantime local efforts appear insignificant. In the pilot project during 1987, local wildlife related employment generated K 70,212 (U.S. $17,553) and a further K 63,600 (U.S. $15,900) was earned toward local community development projects. Although substantial in an otherwise cashless economy (and, more important, locally generated and available for the indefinite future), such an amount is easily denigrated amid discussions of millions from bilateral or multilateral donors.

Based on a continual dialogue with participants, successful community projects must be flexible, creative, and able to adjust and grow based on experience and changing needs. In ADMADE, village meetings play a critical role in soliciting views and criticisms from local residents on wildlife management, both to build a sense of proprietorship and to overcome past antagonism with the NPWS. Resources must be sufficient to generate the income to meet community-identified needs. Unfortunately, ADMADE faces an inherent risk because wildlife is subject to drought, epidemics, and habitat disruption, all of which affect reproductive success. A severe drop in the wildlife resource due to such natural fluxuations could impact the economic sustainability of the program. For conservation to result, there must be a clear link between the revenues and resources. One concern with the larger-scale, multiple-sector, integrated LIRDP project was that commingling of resources from different development activities, such as forestry, farming, and wildlife, may blur the wildlife stewardship and income generating relationship sought in ADMADE.

Faced with massive needs, both human and natural, it is tempting to attack Africa's problems at once and on all fronts, but incremental improvements carry less risk than radical changes or multiple innovations. Incremental improvements also increase prospects for success, which, in turn, instills self-confidence and enthusiasm. Small steps more easily balance the need for change with respect for tradition. Demonstration is critical to allow programs to prove themselves, especially given past antagonism with the government wildlife department. This approach requires patience and a focus on projects where prospects for success are greatest. ADMADE did not start with a depleted GMA, but rather asked questions such as these: Where was the best wildlife resource? The best unit leader? The most committed local chief or provincial governor?

Underlying these experiences is the simple truth that conservationists need to find systems or ways of doing things that will make local communities the owners of their own development and the owners of their own conservation.

Update, 1989–1994

Over the last five years the ADMADE program has been undergoing a dramatic, critical, and risky expansion with the receipt of a major grant from USAID. With the expansion, ADMADE became Zambia national policy, and the community-based approach was applied in all 32 GMAs. However, many successful small-scale projects have failed when institutional strengthening was unable to keep pace with rapid expansion (Yudelman 1991), and ADMADE has not escaped growing pains. Difficulties particularly relate to administration and political weakness in the capital or at NPWS headquarters in Chilanga, while activities in the field have continued to struggle forward slowly.

As was feared, a large influx of donor funds lead the department to neglect the lesson of living within its means. Desperately needed vehicles were obtained from donations, but it proved beyond the ability of wildlife generated revenues to fund replacements. Despite the fact that vehicles were not well maintained and occasionally misappropriated, the training of unit leaders and village scouts has had a demonstrable impact on poaching. The process of decentralization of wildlife decision making has continued, although abuses by some local leaders persist. ADMADE began to experiment with land-use planning through a program linking traditional leaders with the technical and senior staff of NPWS that has broken down some of the barriers of distrust between the rural communities and the central government agency.

Relationships deteriorated between the Zambian-lead NPWS and the white-dominated safari hunting industry due, in part, to the un–business like approach of the government bureaucracy. In 1993 there was a virtual collapse of safari hunting revenue, largely due to adverse publicity from former operators unhappy with the allocation of hunting areas. Zambia neither anticipated nor countered this problem with a marketing effort of its own (although that is now being corrected). ADMADE and the WWF had not analyzed this primarily source of revenue because of the sensitivity of a conservation organization working so closely with an industry abhorred by many of its supporters.

Relationships between the department and the WWF also began to show strains over responsibility for the management of funds and program decisions. The power struggle was exacerbated by the use of expatriate consultants in the place of Zambian personnel and the suggestion of an external evaluation team, concerned with political problems and financial irregularities and perhaps influenced by disgruntled safari operators, that an NGO should take over management of ADMADE, a proposal rejected by the government as a return to a colonial model of conservation.

Equally problematic, weak and politically motivated personnel in the Ministry of Tourism demonstrated the vulnerability of the program and local residence rights unless ADMADE becomes legally instituted. Abuses in the wildlife revolving fund were only addressed when the NPWS's request for an audit was backed by external donors. Civil service salaries made it impossible to compete with the private sector for key financial personnel, and regulations made it difficult to replace wardens whose performance was increasingly being compared unfavorably with that of unit leaders in the field. Recently, the problems of government staff instability turned in ADMADE's favor with the appointment of a knowledgeable and sympathetic permanent secretary in the ministry.

Low-income earnings from safaris combined with revolving fund irregularities brought a cash flow crisis just as the approaching 1993–94 wet season increased prospects of renewed poaching. Showing government's commitment to the program, the Ministry of Finance intervened with a loan to pay village scouts giving the program a reprieve to get its professional house in order. A reorganization is now being proposed giving the department greater freedom to hire and fire, set salaries, and work outside the civil service system. Despite all of its difficulties, ADMADE has maintained the loyalty of local leaders and has slowly been providing schools, clinics, grinding mills, and the provision of jobs from natural resources. Although its future is far from assured, the program's ability to survive and adapt is encouraging.

The Search for "Imperfection"

Does the failure of much of wildlife conservation in Africa simply mirror a wider range of externally promoted initiatives that have contributed to the 1980s' being Africa's "lost decade"? An increasing number of Africans are arguing that, whether borrowed or thrust upon them from the East or from the West, they have been following the wrong model of development, and the continent must now begin to search for particularly African solutions. Similarly, if successful African conservation strategies are to be built, they cannot be based upon a northern mythology of an African "Eden," but on a clear-eyed critique of past failures and a search for African leadership and local experience. The ironic point is that, for too long, externally promoted "ideal" approaches to conservation have been supported at the expense of practical local solutions. On the other hand, ADMADE's record over the last five years reveals the complexity and susceptibility of these local solutions to political manipulation, moribund bureaucracies, and unstable national economies. But with patience and a long-term perspective, the Zambia ADMADE program may yet prove to be an "imperfect" solution, a solution that works in the real world of Africa.

References Cited

Anderson, David, and Richard Grove, eds.
    1987 The Scramble for Eden: Past, Present and Future in African Conservation. *In* Conservation in Africa: People, Policies and Practice. Pp. 1–12. Cambridge: Cambridge University Press.
Asibey, E. O.
    1974 Wildlife as a Source of Protein in Africa South of the Sahara. Biological Conservation 6(1):32–39.
Atkins, S. L.
    Socio-Economic Aspects of the Lupande Game Management Area. *In* Proceedings of the Lupande Development Workshop. D. B. Dalal-Clayton, ed. Pp. 49–56. Zambia: National Parks and Wildlife Service.
Bell, R. H. V.
    1987 Conservation with a Human Face: Conflict and Reconciliation in Africa Land Use Planning. *In* Conservation in Africa: People, Policies and Practice. David Anderson and Richard Grove, eds. Pp. 79–101. Cambridge: Cambridge University Press.
Bell, R. H. V., and J. E. Clarke
    1984 Funding and Financial Control. *In* Conservation and Wildlife Management in Africa. R. H. Bell and E. McShane-Caluzi, eds. Pp. 545–555. Washington, DC: U.S. Peace Corps.
Caldecott, John
    1986 Hunting and Wildlife Management in Sarawak. Kuala Lumpur: World Wildlife Fund and Aysin.
Child, Graham, and W. Nduku
    1985 Wildlife and Human Welfare in Zimbabwe. African Forestry Commission, FAO, FO: AFC/WL: 86/62, October.
Clay, Jason
    1988 Indigenous Peoples and Tropical Forests: Models of Land Use and Management from Latin America. Cultural Survival Quarterly:11–14.
Clay, Jason, ed.
    1985 Parks and People. Cultural Survival Quarterly 9(1).
Cumming, D. H. M.
    1987 A Project Proposal for a Field Study of Multi-Species Indigenous Wildlife Utilization. World Wildlife Fund Project No. 3749, August.

Cumming, D. H. M., and R. D. Taylor
    1989 Identification of Wildlife Utilization Projects for the Department of Wildlife and Parks, Government of Botswana. World Wildlife Fund and Kalahari Conservation Society. Unpublished report.
Deihl, Colin
    1988 Wildlife and the Maasai: The Story of East African Parks. Cultural Survival Quarterly 9(1):37–40.
Development Alternatives
    1989 Regional Natural Resources Management Project: Community-Based Resource Utilization. Project paper submitted to USAID/Zimbabwe, August.
Gardner, J. E., and J. G. Nelson
    1981 National Parks and Native Peoples in Northern Canada, Alaska and Northern Australia. Environmental Conservation 8(3): 207–215.
Hancock, G
    1989 Lords of Poverty. New York: Atlantic Monthly Press.
Hart, J. A., and G. A. Petrides
    1987 A Study of Relationships between Mbuti Hunting Systems and Faunal Resources in Ituri Forest of Zaire. In People and the Tropical Forest. A. Lugo, ed. Pp. 12–14. Washington, DC: U.S. Department of State.
Hitchcock, R.
    1989 Indigenous Peoples and Wildlife Schemes. Kalahari Conservation Society Newsletter 24:10–11.
Hopcraft, D.
    1986 Wildlife Land Use: A Realistic Alternative. In Wildlife/Livestock Interfaces on Rangelands. S. MacMillian, ed. Pp. 993–1101. Nairobi, Kenya: Winrock International.
Hunter, Malcom I.., Robert K. Hitchcock, and Barbara Wyckoff-Baird
    In press Elephants, Women and Biological Diversity in Southern Africa. Conservation Biology.
International Union for the Conservation of Nature
    1981 The Importance and Values of Wild Plants and Animals in Africa. Gland, Switzerland: IUCN/WWF/UNEP.
Kapungwe, E., and D. M. Lewis
    1989 Wildlife Utilization Schemes for Local People in a Wildland: A Proposal Submitted to Africare. October.
Lawson, N.
    1985 Where Whitemen Come to Play. Cultural Survival Quarterly 9(1):54–56.
Lewis, Dale
    1989 A Promise Worth Keeping. Animal Kingdom (May/June):58–63.
Lewis, D., G. Kaweche, and A. Mwenya
    In press Wildlife Conservation Outside Protected Areas: Lessons from an Experiment in Zambia. Conservation Biology.
Luangwa Integrated Resource Development Project
    1987 The Luangwa Integrated Resource Development Project: A Brief Description. Government of Zambia. Unpublished report.
Malama, G.
    1984 Address. In Proceedings of the Lumpanda Development Workshop: An Integrated Approach to Land Use Management in the Luangwe Valley. D. B. Dalal-Clayton, ed. P. 8. Lusaka, Zambia: Government Printer.
Marks, Stewart. A.
    1984 The Imperial Lion: Human Dimensions of Wildlife Management in Central Africa. Boulder, CO: Westview Press.
Matowanyika, Joseph Zans Zvapera
    1989 Cast Out of Eden: Peasants versus Wildlife Policy in Savanna Africa. Alternatives 16(1):30–39.
Muir, Kay
    1988 The Potential Role of Indigenous Resources in the Economic Development of the Arid Environment in Sub-Saharan Africa. Department of Agricultural Economics and Extension Working Papers, AEE, Harare, Zimbabwe, September.

Murphree, M. W., and I. C. Murombedzi
   1987   Wildlife Management Schemes for Zimbabwe's Communal Areas: A Preliminary Survey of Issues and Potential Sites. Centre for Applied Social Science, University of Zimbabwe, May 10.
Ntiamoa-Baidu, Yaa
   1987   West African Wildlife: A Resource in Jeopardy. Unasylua 39(2):27–35.
Owen-Smith, G., and M. Jacobsohn
   1989   Involving a Local Community in Wildlife Conservation. A Pilot Project at Purros, Southwest Kaokoland, SWA/Namibia. Ouagga 27:21–28.
Parkipuny, M. S., and D. J. Berger
   1993   Sustainable Utilization and Management of Resources in the Maasai Rangelands: The Links between Social Justice and Wildlife Conservation. In Voices from Africa. Dale Lewis and Nick Carter, eds. Pp. 113–131. Washington, DC: WWF.
Partridge, W. L.
   1991   Community and Environmental Rehabilitation in Involuntary Resettlement. Paper presented at the 90th annual meeting of the American Anthropological Association.
Poole, Peter
   1989   Developing a Partnership of Indigenous Peoples, Conservationists and Land Use Planners in Latin America. World Bank Policy, Planning and Research Working Papers, pp. 23–26, April.
Redford, K. H., and J. Robinson
   1987   The Game of Choice: Patterns of Indian and Colonist Hunting in the Neotropics. American Anthropologist 89(3):650–667.
Selpapitso, Kgosi
   1988   Legal Aspects: The Traditional View. In Sustainable Wildlife Utilization: The Role of Wildlife Management Areas. P. Hancock, ed. P. 45. Gaborone, Botswana: Kalahari Conservation Society.
Turnbull, Colin M.
   1972   The Mountain People. New York: Simon and Schuster.
Volkman, Toby Alice
   1986   The Hunter-Gatherer Myth in Southern Africa: Preserving Nature or Culture? Cultural Survival Quarterly 10(2):25–32.
Ward, Haskell G
   1989   African Development Reconsidered: New Perspectives from the Continent. New York: Phelps-Stokes Institute.
Wright, Michael
   1988   People-Centered Conservation: An Introduction. World Wildlife Fund Letter (3):1.
Yudelman, Monty
   The Sasakawa–Global 2000 Project in Ghana: An Evaluation. March. FINIDA and the Carter Center. Unpublished report.

Anthropology in Pursuit of Conservation and Development

Michael Painter

Introduction

Anthropologists have long acted as advocates for broadly based local participation in conservation and development activities, and much of our applied work consists of attempting to create a context in which this participation can come about. Anthropology has offered a number of insights into how people utilize the environment, and many anthropologists have played central roles in increasing public awareness of different dimensions of the environmental problems confronting contemporary societies. Discussions concerning land use in tropical forests and other at-risk areas are replete with references to culture, indigenous knowledge systems, and other concepts substantially or totally drawn from anthropological research. Yet interpreting the significance of these concepts for environmental policy and political action has largely been left to nonanthropologists. This is in part because anthropology as a discipline has not defined environmental destruction as an area of concern.

As a result, the application of the findings of anthropological research to environmental problems is inconsistent and anecdotal. For example, research on indigenous environmental knowledge and the functioning of different types of production regimens is widely cited. At the same time, victims of environmental destruction such as smallholding peasants are frequently blamed as destructive agents, despite the fact that the bulk of anthropological literature argues that they normally seek to promote the long-term stability of their production systems and allocate resources rationally toward that end. Anthropologists' response to this apparent paradox is curious. In seemingly interminable conversations among ourselves, we cite the application of our work by others as evidence that what we do is useful. At the same time, we are chagrined and perplexed about the ways in which anthropological research is often used, and we frequently wonder why more individuals and institutions do not come to us for guidance for the definition and solution of environmental problems.[1]

Our difficulties appear to come from different directions. First, anthropology remains largely an academic discipline, uncomfortable with applying the study of humans to the addressing of human problems. We respond to this discomfort either by retreating into the academy and pretending to deal with social issues that transcend concerns about poverty, the environment, and social justice, or we plunge headlong into the service of a wide range of institutions, eschewing the relevance of a large part of anthropological theory to our work and promoting an impoverished series of data gathering techniques as a supposedly value-free applied anthropology.[2] Interestingly, both responses constitute a fundamental denial of the relationship between social science and the social context in which it is practiced.

A second difficulty is that anthropologists mire themselves in self-limiting definitions of what anthropology is. Despite all that has been written about "studying up," research continues to be largely local and historically decontextualized. We are remarkably able to identify adaptive behaviors and strategies in contexts where people are enduring extreme hardship. We are less able to discuss systematically the relationship between the adaptive strategy and the hardship. For example, although many of us write about the marvels of diversified peasant production, we seldom explore how that production may arise from and perpetuate both impoverishment and environmental destruction.

Anthropologists concerned with promoting local participation in the interest of sustainable resource use and economic growth need to base their research and any recommendations arising from it on an explicit understanding of the values we seek to promote through particular types of production regimens and institutional arrangements. Though it may be comforting to pretend that our recommendations are value-free and emerge in a self-evident way from the data we gather, and though it may be necessary for the institutions for which we work to maintain this fiction, to forfeit a critical, anthropological perspective on our own work portends disaster for conservation and development objectives.

Ultimately, we are talking about who wins and who loses in the struggle for the resources upon which many depend to produce and, through their production, to survive and prosper. Environmental destruction and economic underdevelopment are a consequence of relationships between people with diverse and conflicting economic interests. As social scientists, we insist that our recommendations are based on a thorough analysis of empirical data. However, that same critical scientific perspective demands that we recognize that our definition of the problem, which shapes how we collect our data, reflects a set of values and interests we take with us to our work (Martínez Alier 1991).

As we reflect on how anthropologists can increase the impact we have on how people think about and use the environment, it is well that we remember the important place that consideration of the relationship between societies and their physical surroundings have occupied in our theoretical discussions, and the context in which these discussions arose. This may help us understand how anthropological concepts and discussions appear in the popular press and the writings of professionals in other disciplines like ghosts from our past—or skeletons from our closet— and are, in the new context, imbued with significance that takes us by surprise. At the same time, we can see that despite the dissatisfaction we may feel at the recognition that anthropology is accorded in the environmental and development areas, our theoretical heritage provides us with the tools to have a greater impact in shaping general consciousness about these issues and in formulating specific policies, if we decide to use them for those purposes.

Anthropological Perspectives on the Environment

Development and natural resource management have not been central issues in anthropological theory until recently. Resource management discussions did occur in the context of early anthropological efforts to influence government policy. However, anthropologists have not, by and large, examined anthropologically the relationship between the social conflicts of a particular period and resulting efforts to influence state policy, and scholarly anthropological theory, despite the historical strength of interchange between these two areas.

For example, Feit (1986) discusses the case of Frank Gouldsmith Speck, who investigated the relationship between property rights and natural resource husbandry among the Native People of North America around the turn of the century. Speck argued that these populations had defined hunting territories over which particular kin groups had exclusive rights, and within which they pursued resource-use practices intended to prevent the depletion of fish and game. The arguments were part of his effort to counter the claims of the Canadian government—acting in support of groups seeking to take over Native American lands—that in large part rationalized this course on the grounds that Native Americans were profligate resource users because, in the absence of exclusively held property, individuals were not accountable for how they used resources. Speck was successful in influencing state policy regarding the recognition of hunting territories, and his characterization of Native Americans as "natural conservationists" continues to influence popular perceptions of Native Americans. However, his ethnographic materials were removed from this policy context in conceptual discussions within the discipline, where their significance was held to be their utility in refuting the evolutionary models of social change that had been proposed by Lewis Henry Morgan.[3]

The social change implications of state policy toward contemporary Native Americans and their land was not a major issue in the theoretical literature of anthropology at the time, despite the fact that anthropologists participated on all sides of the policy debate. The operative model was that scientific theory was above and independent of temporal matters, and hints that political issues of the day influenced the definition and direction of theoretical discussion undermined its scientific legitimacy.

Nevertheless, political concerns clearly limited anthropological consideration of human interaction with the environment. Some American anthropologists (e.g., Kroeber 1939; Wissler 1938) defined culture areas based on shared traits and related these to geographic regions. The profession was concerned, however, that this not be confused with the environmental determinism of the late 19th century, a belief that sought to justify the economic and political dominance of tropical regions by Europe and the United States on the basis of climatic factors that supposedly favored a more vigorous and industrious population in the temperate latitudes of the northern hemisphere. As a result, the environment was accorded only a passive role in shaping human behavior (Herskovits 1951:159).

Cultural evolutionists such as White (1949) and Childe (1951) also constructed models of social change in which the environment played a passive role. For White, for example, cultural evolution was a function of the ability of a society to capture energy from the environment in order to turn it to productive purposes: the greater the energy captured, the higher the level of cultural evolution. The capacity to capture energy was a function of technological advancement, which in turn depended on the internal organization of a culture and the institutional relationships whereby access to the means of production were regulated. Conceived in a period when natural resources appeared unlimited, White's model considered the total amount of energy captured, not the efficiency of energy use, as the measure of cultural evolution. Nor did he consider how cultures might respond to resource scarcity. For White, "environmental factors may . . . legitimately be considered a constant and as such omitted from our consideration" (1949:199).

It is telling that the major critique of unilinear evolution (Steward 1955) focused on demonstrating that human responses to a given set of conditions are more variable and multifaceted than allowed for by White. However, he did not critically examine basic notions related to the supposed constancy and passivity of the environment, or the lack of natural limits on human capacity to extract energy from it.

Steward's approach turned on the ways in which people use technology to exploit or transform the environment through production. Production was a technical activity mediating adaptation to the environment, and the way people used natural resources to produce was a function of technological advancement. Social organization could change according to the need to manage a particular productive activity at a given level of technology and within the passive constraints imposed by the environment. Reacting to anthropology's tendency to link history to culture history, which was rooted in the idea that culture traits diffused from on population to another, and embodied a rejection of evolutionary models of social change, Steward largely rejected a historical perspective. The anthropological view of history, he felt, trapped one in particularistic explanations of change that precluded linking specific cases to general processes (Steward 1955:78–97, 208–209).

In *The People of Puerto Rico* (1956), Steward et al. argued that changes in production practices and social organization should be explained in terms of environment and technology (1956:15). Yet the volume also contains an extended historical discussion written collectively by the project's research staff (Steward et al. 1956:29–89) that reflects the limitations they found in cultural ecology's ability to interpret adequately the data collected by the study (Roseberry 1978:31). The historical discussion addresses Steward's concern to avoid being trapped in particularistic explanations by explaining variations in Puerto Rican subcultures in terms of local responses to the demands of international commodity markets and the resulting reorganization of production relations (Steward et al. 1956:32). The introduction of historical explanation also begins to overcome weaknesses in the formulation of cultural ecology. For example, cultural ecology's interpretive power was limited by the focus on the immediate users of a technology and the environment, which made it difficult to place local situations in a broader context of change. Similarly, the lack of a historical perspective made it difficult for cultural ecology to explain the process leading to a particular local situation.

At the same time, the limitations of the study reflect theoretical and methodological issues with which anthropologists continue to grapple. For example, the study defined agricultural production in a narrow, empirical way, but did not discuss its place within broad patterns of capital accumulation affecting Puerto Rico, nor did it explicitly discuss capitalism itself. Thus, phenomena such as migration to the U.S. mainland, which were locally important, but which owed their origin and perpetuation to forces that were at work internationally, were not considered (Wolf 1978, 1990:589).

Nonetheless, *The People of Puerto Rico* demonstrated that patterns of production and resource use are more than the outcome of the application of a particular level of technology to a set of environmental constraints. The study provides a point of departure for understanding how the definition of what natural resources are important in a particular time and place, how and to what end they are exploited, and the overall relationship of the population to the physical environment flow from the social relations among people.

During the 1960s, American economic anthropologists were heavily involved in a debate over the appropriate way to conceptualize how goods are distributed in non-Western societies. Ultimately, they became mired in disagreements over the suitability of subjecting these societies to formal economic analysis. Responding to critiques from Marxist scholars (e.g., Godelier 1977) and from human ecologists (e.g., Vayda 1967, 1969), economic anthropologists began to focus on production rather than distribution as the starting point of their analysis. Because it arose out of concern about the exchange relations among people, the approach adopted by some economic anthropologists lent itself more easily to treating production as a social activity rather than as a purely technical one (Cook 1973, 1974).

By examining resource use in terms of its socially defined purpose, anthropologists demonstrated that who controls access to natural resources and the institutional arrangements through which that control is mediated determines the sorts of resource management practices that will be followed, within the constraints of the physical environment. Based on this insight, anthropology began to view natural resource management issues in a new way. Rather than treat the environment as a passive entity that imposes broad limits on human activity, some anthropologists shifted their focus to the dynamic relationship between human productive activity and the physical resource base. The nature of this activity is shaped by the distribution of access to productive resources and the nature of the institutional arrangements that mediate that access.

Wolf (1972) coined the term "political ecology" to describe an approach to the study of human resource use that centers on who controls access to key resources and to what end their control causes production to be dedicated. Since that time, a number of anthropologists have applied the notion—sometimes explicitly and sometimes implicitly—very productively to the problem of how human productive activity contributes to environmental destruction. Hjort (1982), for example, used the term in his critique of ecological models of pastoral land use. Hjort notes that traditional ecological discussions attempt to treat human populations the same way they would treat other animals in an ecosystem. This reduces patterns of resource use by humans to a technical question and imparts an illusory, normative quality to those patterns. As Hjort points out, defining a normal state for contemporary African pastoral production systems is an ideologically charged issue because the "normal" condition is one of extreme seasonal and annual fluctuations in vegetation and constant alternation between periods of scarcity and abundance (1982:15). Similarly, a purely ecological approach does not take into account the chronic land grabbing by those with wealth and political power, which has also been part of the normal state of affairs for contemporary pastoral populations, reducing both the geographic area and biological diversity to which such peoples have access (Hjort 1982:20).

Schmink and Wood (1987) propose elements that should be part of a political ecology approach to understanding the processes of settlement and environmental destruction affecting the Amazonian lowlands of South America. These include (1) the degree to which production is oriented toward simple reproduction or capital accumulation; (2) the class structure of the society to which the region in question belongs and the lines of conflict over access to productive resources; (3) the extent and kinds of market relations in which producers are involved, and the mechanisms whereby production beyond that needed to satisfy consumption requirements is extracted as surplus; (4) the role of the state in defining and executing policies that favor the interests of certain classes of resource users over others; (5) international interests, such as donor agencies or private investors, that may support particular patterns of resource use; and (6) the ideology that orients resource use—for example, the position that rapid economic growth is the best way to address social and environmental problems—and what groups benefit from that ideology.

A number of authors have profitably developed the political ecology line of argument in several world areas. These include Collins (1986, 1987, 1988), Durham

(1979), Edelman (1985, 1992), Little (1985), Little and Brokensha (1987), and Stonich (1989, 1992, 1993), among others. They have done much to illustrate how patterns of environmental destruction are associated with particular social processes, including chronic labor scarcity arising from the dependence of rural families on off-farm employment, indebtedness, declining incomes, and class and ethnic conflict. They have also used innovative methodologies to link local-level processes with patterns of capital accumulation occurring at regional, national, and international levels.

Environmental Destruction in Lowland Bolivia Julian

In 1984, the Institute for Development Anthropology (IDA) began a long-term study of the relationships between particular production relations, processes of accumulation, and patterns of impoverishment and environmental destruction. The initial phase of this research was a study of settlement in the eastern Bolivian lowlands, focusing on the San Julian colonization project. The results of this research illustrate the importance of contextualizing local-level data collection within broader historical processes in order to assess the environmental implications of the development efforts being conducted.[4]

Background to Lowland Settlement

As in other South American countries, lowland settlement has long been seen by nationals and foreigners alike as a means of solving social and economic ills in Bolivia's upland areas. This became important in the wake of the 1952 revolution. In upland areas the government undertook an agrarian reform that redistributed large areas of land to smallholders. It also assigned a low priority to the development of smallholder agriculture and pursued policies, such as heavy reliance on imported foodstuffs to satisfy urban demand, that were hostile to smallholder interests (Frederick 1977). This accelerated the deterioration of rural living conditions, resulting in massive redistribution of the population. Many families turned to long-term or permanent migration outside of Bolivia (Balán and Dandler 1986; Whiteford 1981); others migrated to Bolivia's cities, primarily to La Paz, Cochabamba, and Santa Cruz (Pérez-Crespo 1991); still others migrated to lowland areas as agricultural laborers on commercial farms or as settlers (Blanes 1984; Stearman 1976, 1985). Most families who remained in upland rural areas earned more than half of their income off-farm.

Following the revolution, the lowland areas around Santa Cruz were the object of an intensive effort to promote the development of large-scale commercial agriculture. Little land was actually expropriated in the lowlands, although the threat of expropriation under the agrarian reform legislation encouraged landowners to regularize their titles and take advantage of the development incentives being offered (Heath 1969:291–295; Ybarnegaray de Paz 1992:88). Largely funded by international donors, these included provisions for agricultural credit, subsidized access to heavy equipment for land clearing, the construction of infrastructural and agro processing facilities, and agricultural research support. Santa Cruz became a continental center for the production of a number of agricultural commodities, including cotton, sugar cane, peanuts, and soybeans (e.g., Gill 1987; Heath 1969; Ybarnegaray de Paz 1992).

The expansion of commercial agriculture in Santa Cruz depended heavily on the cheap labor made available through the impoverishment of rural families in upland areas. In 1976, for example, agricultural employment in Santa Cruz fluctuated between 18,000 people in February and 95,000 people in August, the month when the greatest amount of land clearing is done (Rivière d'Arc 1980:158). Despite the large number of people who sought work in Santa Cruz, producers consistently found the labor supply inadequate and sought state help in securing workers. Workers were on occasion recruited against their will, and troops were called out to provide labor for the 1973 cotton harvest (Gill 1987:69–70).

Many of the people who originally went to Santa Cruz as workers brought their families and remained in areas designated for settlement around the fringes of the commercial agricultural areas. In part to create a resident force for the commercial farms, and in part to alleviate the social and economic pressures afflicting upland areas, the state began to encourage smallholder settlement in Santa Cruz. Most of these efforts yielded disappointing results. Though some settlers were better able to feed themselves and their families through farming than had been the case in their home areas, they remained poor and tended to become poorer. Consequently, they faced a number of pressures to use land increasingly extensively and destructively.

Begun in 1972, the San Julian project attempted to learn from these negative experiences and promote a style of settlement that would result in self-sustaining communities that could be expected to grow and prosper over the long term. The approach encouraged settler participation in planning and implementing activities carried out in San Julian. Within the confines of the project area, the project succeeded admirably in solving problems that had been the bane of settlement efforts elsewhere in the region. However, because it did not improve the conditions under which settlers participated in the regional economy, major pressures promoting destructive patterns of land use remained in force.

Pressures for Deforestation and Land Degradation

In San Julian, as in much of Santa Cruz, smallholder agriculture was based on the cultivation of rice and corn.[5] Although sugar cane, cotton, and soybeans occupied a large part of the department's agricultural land, they tended to be grown on large commercial enterprises. For reasons to be described below, commercial enterprises were not heavily involved in rice and corn production.

As smallholder crops, rice and corn played a dual role, providing the basic elements, along with yucca (cassava), of the settler diet and serving as the most important cash crops. Indeed, concerned with establishing settlers' food security as quickly as possible, the San Julian project promoted these as the central elements in their farming systems. However, since the early 1960s, Santa Cruz produced more rice than the national market could absorb (Heath 1969:289–299; Hiraoka 1980). In 1984, Santa Cruz farmers grew about 150,000 tons of rice, only about 105,000 of which was absorbed by national demand. The situation with corn was similar, with the department producing about 150,000 metric tons and national demand absorbing only about 100,000 metric tons. Efforts to expand the market for rice and corn confronted a series of technical and economic obstacles that made this an unlikely option for the foreseeable future (CORDECRUZ 1984a, 1984b; Farrington 1984:18–19).

As a result, income for settlers declined as the price they received for their corn and rice declined in relation to production costs. The decline was most dramatic for the poorest producers, who had little access to mechanized equipment and relied most heavily on family and wage labor, as the price of corn and rice lost ground more rapidly to labor costs than it did to the costs associated with mechanized production. Poorer settlers were also hit particularly hard by transport costs, which were generally high in the settlement area and went up precipitously if the trucks had to leave the main road.

Settlers responded to declining income in several ways, including (1) seeking to incorporate new, high-value crops into their production system; (2) setting up small retail stores to supplement agricultural income; and (3) attempting to withdraw into production focused on satisfying family consumption needs as fully as possible without relying on corn and rice sales. In each case, these efforts proved to be stopgap measures at best.

Ultimately, declining incomes generated pressures for farmers to increase their revenues by increasing the amount of product they offer for sale. In many cases, this was accomplished by intensifying family labor inputs. However, the lower prices resulting from increased supply quickly brought most families to the limit of their ability to get by through working harder. At that point, the only remaining option was to become more extensive in land use, clearing large areas of forest with no intention of keeping it in production, and moving to a new area as soon as yields began to fall.

The pressures for settlers to use land extensively could be reduced or eliminated through improvements in areas that influence the nature of their relations with the regional economy. Road construction and improvement, the encouragement of competition in transport, the establishment of bulking, storage and rudimentary agroprocessing facilities, and the promotion of collective production and marketing strategies through the various grassroots organizations in which settlers participate would improve the conditions under which corn and rice are sold and permit the immediate expansion of the variety of crops settlers offer for sale. Research and extension directed at smallholder production and marketing problems would have additional medium- and long-term impacts.

However, development resources go into three major areas: large-scale production of industrial export crops, lumber, and oil. Within these areas, some are openly hostile to promoting agricultural development among the settlers; others simply do not consider it a priority concern. The net result is that although laborers and settlers were an essential element of the economic growth of Santa Cruz, they are excluded from consideration in the allocation of development resources. Efforts on the settlers' part to organize and participate in regional politics in order to represent their interests are met with hostility and violence (e.g., APDHB 1984; Painter 1988). Thus, the solution to environmental destruction in the settlement area lies not in altering the relationship between settlers and the land, but in altering the relationships between settlers and other social classes competing for economic resources.

Conclusion: Implications for Conservation and Development

The experience of San Julian underscores the importance of conducting analyses that place resource competition in a historical context in order to identify the social and economic underpinnings of environmental destruction. It is not enough to promote new technology and local participation. In areas such as eastern Bolivia, where smallholding settlers, modern agribusiness, and lumber and oil companies exist side by side, technology and participation are hollow notions unless we specify technology to be used by whom and for what purpose, participation by whom, and under what institutional arrangements. Only then do the issues of access to resources underlying environmental destruction and underdevelopment emerge, and only then can we examine the kind of world we wish to build through the exploitation and/or conservation of natural resources.

Redclift notes:

[T]he environment, whatever its geographic location, is socially constructed. The environment used by ramblers in the English Peak District, or hunters and gatherers in the Brazilian Amazon, is not merely *located* in different places; it means different things to those who use it. The environment is transformed by economic growth in a material sense but it is also continually transformed existentially, although we—the environment users—often remain unconscious of the fact. [1987:3, emphasis in original]

This is certainly applicable to the case of eastern Bolivia; the significance of the environment varies fundamentally according to the opportunities to use it that one's class position offers.

The environment, however, is more than a socially constructed category, for all of the productive activities discussed in this article are ultimately limited by the physical capacity of the natural resource base to sustain them (Benton 1989; Grundmann 1991). It is in this intersection of the social construction of the environment by humans from different social classes, involved in diverse labor processes and standing in distinct relationships to processes of accumulation and impoverishment, and the physical limits that the environment places on human production where anthropology is uniquely qualified to inform about the conservation and development options confronting us at particular times and places. What is required is that we place our research explicitly in the context of the political debate over these options. To the extent we do this, not only will anthropological research be widely cited, but anthropologists' interpretations of what the research means will be increasingly sought and valued.

Notes

Acknowledgments. Support for writing this article was provided by the Cooperative Agreement on Settlement and Resource Systems Analysis of Clark University, the Institute for Development Anthropology, and the U.S. Agency for International Development. However, the views expressed are those of the author and should not be attributed to any of the institutions mentioned or individuals acting on their behalf.

1. For example, the *Anthropology Newsletter* complains:

Two cover stories on the environment (*Time* Magazine, "Torching the Amazon: Can the Rainforest Be Saved," September 18, 1989; and *Scientific American*, Special Issue: "Managing Planet Earth," September 1989) are peppered with anthropology-relevant words and phrases: "culture shift," "traditional culture," "pronatalist culture," "peasants," "Neolithic," "humankind," "desertification," and so on. Yet *anthropologists* are neither quoted or cited anywhere in the magazines. [1989:2, emphasis in original]

2. See, on the one hand, Ehler's (1991) discussion of the lack of attention given to violence and oppression in articles dealing with Central America that have appeared in leading anthropological journals. On the other hand, Gow (1993:392) notes that anthropologists who do attempt to address human problems in the development arena often do not avail themselves of the powerful critical analysis tools that anthropology offers.

3. Commenting on a collection of articles addressing the social responsibility of anthropologists, Beals (1968:407–408) traces anthropological reluctance to address pressing social issues to the late 19th and early 20th centuries. Anthropologists' recommendations regarding legislation and policy concerning Native People in the United States sometimes led to unexpected and very negative consequences when followed, creating the widespread sense that our theoretical and analytical tools were inadequate for addressing complex social problems. It is worth recalling, with Beals, that anthropologists have been very involved individually in attempting to address social issues.

4. Subsequent research has examined the socioeconomic and environmental processes that drive migration from upland areas and the organization of production in Bolivia's Chapare region. Funding for the research has been provided by the Cooperative Agreement on Settlement and Natural Resource Systems Analysis of the Institute for Development Anthropology, Clark University, and the U.S. Agency for International Development, and by the United Nations Development Fund for Women. The views expressed here are the author's and do not necessarily reflect the positions of any of these institutions. Those interested in more detailed accounts of the San Julian research should see Painter 1987, Painter and Partridge 1989, Painter et al. 1984, and Pérez-Crespo 1987.

5. The information presented in highly summarized form here is discussed in detail in Painter 1987.

References Cited

Anthropology Newsletter
    1989   Conspicuously Absent. Anthropology Newsletter 30(8):2.
APDHD
    1984   San Julián: Bloqueos campesinos y camioneros. La Paz: Asamblea Permanente de Derechos Humanos de Bolivia.
Balán, J., and J. Dandler
    1986   Marriage Process and Household Formation: The Impact of Migration on a Peasant Society. Report prepared for the Population Council.
Beals, R.
    1968   Comment. Current Anthropology 9(5):407–408.
Benton, T.
    1989   Marxism and Natural Limits: An Ecological Critique and Reconstruction. New Left Review 178:51–86.
Blanes, J.
    1984   De los valles al Chapare. Cochabamba: CERES.
Childe, V. G.
    1951   Man Makes Himself. New York: New American Library.
Collins, J. L.
    1986   Smallholder Settlement of Tropical South America: The Social Causes of Ecological Destruction. Human Organization 45(1):1–10.
    1987   Labor Scarcity and Ecological Change. In Lands at Risk in the Third World: Local-Level Perspectives. P. D. Little and M. M. Horowitz, eds. Pp.17–37. Boulder, CO: Westview Press.
    1988   Unseasonal Migrations: The Effects of Rural Labor Scarcity in Peru. Princeton: Princeton University Press.
Cook, S.
    1973   Production, Ecology, and Economic Anthropology: Notes toward an Integrated Frame of Reference. Social Science Information 12(1):25–52.
    1974   Economic Anthropology: Problems in Theory, Method, and Analysis. In Handbook of Social and Cultural Anthropology. J. J. Honigmann, ed. Pp. 795–860. Chicago: Rand McNally.
CORDECRUZ
    1984a   Arroz boliviano. Boletín Informativo Agropecuario, Departamento de Comercialización Agropecuaria, Corporación Regional de Desarrollo de Santa Cruz 2:1–2.
    1984b   Maíz: Enfrentando una sobreproducción. Boletín Informativo Agropecuario, Departamento de Comercialización Agropecuaria, Corporación Regional de Desarrollo de Santa Cruz 1:3–4.

Durham, W. H.
    1979   Scarcity and Survival in Central America: The Ecological Origins of the Soccer War.
        Stanford, CA: Stanford University Press.
Edelman, M.
    1985   Extensive Land Use and the Logic of the Latifundio: A Case Study in Guanacaste
        Province, Costa Rica. Human Ecology 13(2):153–185.
    1992   The Logic of the Latifundio: The Large Estates of Northwestern Costa Rica since
        the Late Nineteenth Century. Stanford: Stanford University Press.
Ehlers, Tracy Bachrach
    1991   Central America in the 1980s: Political Crisis and the Social Responsibility of
        Anthropologists. Latin American Research Review 26(3):141–158.
Farrington, J.
    1984   El desarrollo de los precios mayoristas de los principales productos agropecuarios
        en Santa Cruz. Boletín Informativo Agropecuario, Departamento de Comercialización
        Agropecuaria, Corporación Regional de Desarrollo de Santa Cruz 1(7):18–27.
Feit, H. A.
    1986   Anthropologists and the State: The Relationship between Social Policy Advocacy
        and Academic Practice in the History of the Algonquin Hunting Territory Debate,
        1910–50. Paper presented at the annual meeting of the American Anthropological
        Association, Philadelphia, December 3–7.
Frederick, R. G.
    1977   United States Aid to Bolivia, 1953–1972. Ph.D. dissertation. Government and Politics
        Department, University of Maryland, College Park.
Gill, L.
    1987   Peasants, Entrepreneurs, and Social Change: Frontier Development in Lowland
        Bolivia. Boulder, CO: Westview Press.
Godelier, M.
    1977   Anthropology and Economics. In Perspectives in Marxist Anthropology, by M.
        Godelier. Pp. 15–63. Cambridge: Cambridge University Press.
Gow, D. D.
    1993   Doubly Damned: Dealing with Power and Praxis in Development Anthropology.
        Human Organization 52(4):380–397.
Grundmann, R.
    1991   The Ecological Challenge to Marxism. New Left Review 187:103–120.
Heath, D. B.
    1969   Land Reform and Social Revolution in the Bolivian Oriente. In Land Reform and
        Social Revolution in Bolivia, by D. B. Heath, C. J. Erasmus, and H. C. Beuchler. Pp.
        241–363. New York: Praeger Publishers.
Herskovits, M.
    1951   Man and His Works. New York: Alfred A. Knopf.
Hiraoka, M.
    1980   Settlement and Development in the Upper Amazon: The East Bolivian Example.
        Journal of Developing Areas 14:327–347.
Hjort, A.
    1982   A Critique of "Ecological" Models of Pastoral Land Use. Nomadic Peoples 10:11–27.
Kroeber, A. L.
    1939   Cultural and Natural Areas of Native North America. University of California Publi-
        cations in American Archaeology and Ethnology 48:1–242.
Little, P. D.
    1985   Absentee Herders and Part-Time Pastoralists: The Political Economy of Resource
        Use in Northern Kenya. Human Ecology 13(2):131–151.
Little, P. D., and D. W. Brokensha
    1987   Local Institutions, Tenure, and Resource Management in East Africa. In Conserva-
        tion in Africa. D. Anderson and R. Grove, eds. Pp. 193–209. Cambridge: Cambridge
        University Press.
Martínez Alier, J.
    1991   Ecology and the Poor: A Neglected Dimension of Latin American History. Journal
        of Latin American Studies 23(3):621–639.

Painter, M.
   1987 Unequal Exchange: The Dynamics of Settler Impoverishment and Environmental
        Destruction in Lowland Bolivia. *In* Lands at Risk in the Third World: Local-Level Perspec-
        tives. P. D. Little and M. M. Horowitz, eds. Pp. 164–191. Boulder, CO: Westview Press.
   1988 Competition and Conflict in the Bolivian Lowlands: Ethnicity and Social Class
        Formation. Paper presented to the invited session "Conceptualizing Inequality: Class,
        Gender, and Ethnicity in the Andes," annual meeting of the American Anthropological
        Association, Phoenix, AZ.
Painter, M., and W. L. Partridge
   1989 Lowland Settlement in San Julian, Bolivia—Project Success and Regional Under-
        development. *In* The Human Ecology of Tropical Land Settlement in Latin America. D.
        A. Schumann and W. L. Partridge, eds. Pp. 340–377. Boulder, CO: Westview Press.
Painter, M., C. A. Pérez-Crespo, M. Llanos Albornoz, S. Hamilton, and W. Partridge
   1984 New Lands Settlement and Regional Development: The Case of San Julian, Bolivia.
        Working Paper 15. Binghamton: Institute for Development Anthropology.
Pérez-Crespo, C. A.
   1987 San Julián: Balance y desafíos. *In* Desarrollo Amazónico: Una perspectiva latino
        americana. Centro de Investigación y Promoción Amazónica and Instituto Andino de
        Estudios de Población, eds. Pp. 185–207. Lima: CIPA and INANDEP.
   1991 Why Do People Migrate? Internal Migration and the Pattern of Capital Accumulation
        in Bolivia. Working Paper 74. Binghamton: Institute for Development Anthropology.
Redclift, M. R.
   1987 Sustainable Development: Exploring the Contradictions. London: Methuen.
Rivière d'Arc, H.
   1980 Public and Private Agricultural Policies in Santa Cruz (Bolivia). *In* Land, People, and
        Planning in Contemporary Amazonia. F. Barbira-Scazzocchio, ed. Pp. 154–161. Cam-
        bridge: Cambridge University Centre of Latin American Studies.
Roseberry, W.
   1978 Historical Materialism and *The People of Puerto Rico*. Revista/Review Interameri-
        cana 8(1):26–36.
Schmink, M., and C. H. Wood.
   1987 The "Political Ecology" of Amazonia. *In* Lands at Risk in the Third World: Local-Level
        Perspectives. P. D. Little and M. M. Horowitz, eds. Pp. 38–57. Boulder, CO: Westview
        Press.
Stearman, A. M.
   1976 The Highland Migrant in Lowland Bolivia: Regional Migration and the Department
        of Santa Cruz. Ph.D. dissertation. Anthropology Department, University of Florida,
        Gainesville.
   1985 Camba and Kolla: Migration and Development in Santa Cruz, Bolivia. Gainesville:
        University Presses of Florida.
Steward, J. H.
   1955 Theory of Culture Change: The Methodology of Multilinear Evolution. Urbana:
        University of Illinois Press.
Steward, J. H., R. A. Manners, E. R. Wolf, E. Padilla Seda, S. W. Mintz, and R. L. Scheele
   1956 The People of Puerto Rico. Urbana: University of Illinois Press.
Stonich, S. C.
   1989 The Dynamics of Social Processes and Environmental Destruction: A Central
        American Case Study. Population and Development Review 15(2):269–296.
   1992 Struggling with Honduran Poverty: The Environmental Consequences of Natural
        Resource Based Development and Rural Transformation. World Development
        20(3):385–399.
   1993 I Am Destroying the Land: The Political Ecology of Poverty and Environmental
        Destruction in Honduras. Boulder, CO: Westview Press.
Vayda, A. P.
   1967 On the Anthropological Study of Economics. Journal of Economic Issues 1:86–90.
   1969 Comments on Dalton's Review Article. Current Anthropology 10:95.
White, L.
   1949 The Science of Culture: A Study of Man and Civilization. New York: Grove Press.

Whiteford, M.
    1981   Workers from the North: Plantations, Bolivian Labor, and the City in Northwest Argentina. Austin: University of Texas Press.
Wissler, C.
    1938   The American Indian: An Introduction to the Anthropology of the New World. 3rd ed. New York: Oxford University Press.
Wolf, E.
    1972   Ownership and Political Ecology. Anthropological Quarterly 45:201–205.
    1978   Remarks on *The People of Puerto Rico*. Revista/Review Interamericana 8(1):17–25.
    1990   Distinguished Lecture: Facing Power—Old Insights, New Questions. American Anthropologist 90:586–596.
Ybarnegaray de Paz, R.
    1992   El espíritu del capitalismo y la agricultura cruceña. La Paz: Centro para el Estudio de las Relaciones Internacionales y el Desarrollo (CERID).

Anthropology, the Environment, and the Third World: Principles, Power, and Practice

David D. Gow

There is a cyclone fence between
ourselves and the slaughter and behind it
we hover in a calm protected world like
netted fish, exactly like netted fish.
It is either the beginning or the end
of the world, and the choice is ourselves
or nothing.

[Forché 1981:59]

How will we feel the end of nature? In many ways, I suspect . . . as with the death of a person, there is more than simply loss, a hole opening up. There are new relationships developing, and strains and twists in old relationships.

[McKibben 1990:65]

Introduction

In order to survive as a profession, anthropology has moved with the times, often reflecting the priorities and agendas established in the North that pay the bills to finance activities determined to be in the best interests of the South. Although a case can be made for the mitigating effects of anthropological knowledge in the field of environmental management and rural development in the third world, particularly activities financed by external actors, a strong argument can also be presented for the opposite—that anthropology has contributed little and, in fact, may have helped to perpetuate the very inequities that such activities are designed, in theory at least, to address.[1] Rather than rehash the well-known arguments for both positions, I wish to steer a middle ground—at least for the moment—and proceed on the basis of certain assumptions that can perhaps point us, as practicing anthropologists, in more challenging directions for the 1990s, the Decade of the Environment.

A first assumption is that the current global interest in the environment will continue, as will the political commitment and the supporting financial resources. The Earth Summit in 1992 provided a reasonable indication of where environmental issues are headed, whether in terms of climate change, biodiversity, or tropical forests. A second assumption relates to the changing nature of this discourse between the North and the South and the extent to which the former can continue to set the environmental agenda, an issue addressed at Rio. On the national level, there is more than a passing ring of paternalism and condescension to environmental planning supported by external donors, whether in the guise of Tropical Forest Action Plans, National Environmental Action Plans, or, most recently, a Global Environmental Facility.[2] A third assumption concerns institutional credibility—that those who finance and those who implement are serious when they reiterate the

46

rhetoric of popular participation, particularly those on the environmental side of the equation:

> [W]hen all is said and done, conservation is about people. It is about the balance that must be struck between humans and nature and between generations. And, if it is to be effective in the developing world, it must address both the needs of nature and the needs of the poor and the dispossessed who ironically share their rural frontier with the Earth's biological wealth. [Wright 1991:4]

On the basis of these assumptions, I wish to argue that if anthropology is to be seriously engaged in solving the environmental equation, there is a need to establish some common ground in the form of basic principles about environmental management and sustainable development that we can support. I wish to demonstrate that anthropology does have some substantive contributions to make to environmental issues, although they leave much to be desired. Finally, I wish to suggest some challenges for anthropology in the 1990s.

Principles for Environmental Management

In much of the third world, conservation for the sake of conservation—environmental fundamentalism—has become an anachronism. There is an increasing awareness and acceptance that if the natural resource base is to be sustained, it must be done so in some productive manner that benefits the local population. Respect for the environment, particularly natural resources, must be accompanied by respect for human resources (Gow 1992a). One way to work toward this is to specify what we mean by environmental management, defined as the field that seeks to balance human demands upon the earth's natural resource base with the natural environment's ability to meet these demands on a sustainable basis. Not surprisingly, given the breadth and complexity of the topic, it covers a wide spectrum, ranging from environmental assessment through various permutations on the theme of resource management. Here, I am concerned primarily with resource management because environmental assessments have distinct limitations and are often seen as an imposition, a requirement mandated by international institutions, the purpose of which often appears to be pro-environment but antipeople, designed more to hinder than to facilitate and sustain.[3]

Resource management, in contrast, pursues a more balanced approach. The underpinning of resource management is the incorporation of all types of capital and resources—biophysical, human, infrastructural, and monetary—into calculations of national accounts, productivity, and policies for development and investment planning (Colby 1990). As the depletion of natural resources is a matter of increasing concern, the interdependence and multiple values of various resources are taken into greater account. For example, increasing attention is being paid to the role of forests and trees in watersheds because they affect hydropower, soil fertility and agricultural productivity, climate regulation, and even fisheries productivity.

The assumption underlying this approach is no longer how to safeguard the environment, but rather how to make more productive use, indeed expanded use, of the natural resource base in order to further human welfare.[4] The key questions to be addressed include the following:

> What sorts of resource depletion are under way? What is their scale and scope? What types and levels of depletion are significant, serious, critical, or intolerable? Are there

thresholds of irreversible injury? What are some forms of interconnection among various categories of resource depletion? How and to what extent does resource depletion generate adverse impacts on sectors, such as food production and public health, that are important for economic development? [Myers 1989:58]

Nor is there any intrinsic reason why concern for the environment, people, and sustainable development must be predominantly technical. The fundamental premise of much mainstream thinking about sustainable development is a two-way link between poverty and environmental degradation, when reality is really much more complicated because both have deep and complex causes. A convincing argument can be made that differential access to resources and the resulting affluence, in the form of overconsumption, may be linked much more directly to environmental degradation than is poverty per se, in either the North or the South (Lele 1991).

Differentiating between ecological and social sustainability is a first step toward clarifying some of the discussion. At some minimal level, ecological sustainability should mean that the local population does not degrade its natural resource base, at least not irretrievably, but rather conserves or even improves it (Gow 1992b). For example, the definition favored by the Brundtland Commission refers to the maintenance or enhancement of resource productivity on a long-term basis that meets the needs of the present without compromising the ability of further generations to meet their own needs.

It should be noted, however, that this definition does accept that ultimate limits exist (World Commission on Environment and Development 1987). In this context, it is necessary to go beyond the rather simplistic notion of sustainable yield and consider the dynamic behavior of the resource in question, particularly uncertainties about environmental conditions, the interactions between resources and activities, and between different uses or features of the same resource (Lele 1991). Equally important from the perspective of social sustainability is the fact that the Bruntland Commission regards sustainable development as a policy objective, quite properly the end point of development aspirations.[5]

The more enlightened conservation and environmental literature of the past decade has consistently argued for a more holistic approach. Based on an in-depth review of this literature, Gardner (1989) has presented a series of substantive principles to promote sustainable development, some of which should have an intrinsic appeal to anthropologists.

The first of these principles is the satisfaction of human needs, based on the precept that ecosystem conservation depends upon the sustenance of the human culture that determines the way resources are used. In third world countries, this depends in turn upon economic development.

The second is the maintenance of ecological integrity, which encompasses the three goals of the World Conservation Strategy dealing with living resource conservation: maintaining essential ecological processes and life support systems; preserving genetic diversity; and ensuring the long-term utilization of species and ecosystems (International Union for the Conservation of Nature 1980).

The third substantive principle for sustainable development is achieving equity and social justice—a prominent theme at the World Conservation Strategy Conference in 1986. The importance of establishing equity is based on the fact that historical patterns of resource use repeatedly demonstrate the importance of commonality of interest and egalitarianism in environmentally prudent behavior (Jacobs et al. 1987). Equity between generations springs from the ethic of maintaining diversity of opportunity for future generations—as indicated by policies and programs that take into consideration their needs and expectations.

The final principle is provision for social self-determination and cultural diversity. A fair distribution of power in decision making can promote this, as can an appreciation, awareness, and incorporation of local knowledge and values (Gardner 1989). Reflecting these concerns, the International Union for the Conservation of Nature (IUCN), the United Nations Environmental Program (UNEP), and the World Wildlife Fund (WWF) have issued a call for a world ethic of sustainability that emphasizes the symbiotic relationships that people have with nature and the values underlying such relationships. The key value proposed, as least from the perspective of the social, is stated as follows:

> [C]ommunities need effective control over their own affairs, including secure access to resources and an equitable share in managing them; the right to participate in decisions; and education and training. Land tenure, other property rights, and the power to decide within the community on allocation of share resources are crucial. [International Union for the Conservation of Nature et al. 1991:9–10]

Some of these principles are embodied in the Global Biodiversity Strategy, particularly the emphasis on cultural diversity, public participation, and respect for basic human rights. In fact, the strategy forcefully advocates for local empowerment, and one of the major objectives is to "correct imbalances in the control of land and resources that cause biodiversity loss and develop new resource management partnerships between government and local communities" (World Resources Institute et al. 1992:80).

In sum, principles that embrace such broad values as the satisfaction of human needs, the achievement of equity and social justice, the furtherance of self-determination, and the promotion of cultural diversity all fall within the realm—in theory, at least—of what is understood to be anthropology (Gow 1992a). Or do they?

Anthropology, Development, and the Environment

Social Analysis

The roles played by anthropologists vary considerably—as do their contributions. Bennett (1988) provides a broad definition that incorporates several overlapping roles. First is the *research facilitation* role, where the anthropologist is responsible for doing background research to show how the objectives of a particular project can be met with minimal social disruption. Second is the *participant employee* role, where the anthropologist is a member of a project management team, responsible for actually implementing the project. Third is the *research evaluation* role, where the anthropologist may be hired to evaluate the results of the project either during implementation or after it is terminated. Finally, there is the more traditional academic role, that of *long-term researcher,* which addresses a variety of developmental priorities jointly established by the sponsors and the researchers (Little 1991).

Much of the work undertaken by anthropologists falls in the research facilitation role, where the anthropologist is responsible for collecting, analyzing, and presenting information on the feasibility of some developmental activity, often in the course of conducting a social analysis. Many anthropologists, I suspect, cut their developmental teeth on social analysis, which, at least in theory, describes and analyzes the real or potential effects of planned interventions upon specific groups of people. Although there is a general belief that such analysis has contributed to the design and implementation of better rural development projects in the third world, the evidence is fragmentary and dispersed (Burdge 1990; Kottak 1991).

Gow et al. (1989) reviewed 15 years of experience of social analysis conducted by the United States Agency for International Development (USAID), one of the pioneers in this field, and their conclusions did little to change this perception. They examined three sources: the literature written by social scientists and development practitioners; the views of those familiar with social analysis; and the documents produced by rural development projects. Though the three perspectives did support the general belief that social analysis has contributed to the design and implementation of better rural development projects, certain strong criticisms also emerged. First, social analysts were not sufficiently critical and rigorous. Many social analysts do not question the basic goals, assumptions, or logic of the project (Morgan 1985). Within a development bureaucracy, too critical a stance has little impact—particularly if the criticisms come from consultants ultimately dependent on that selfsame bureaucracy for their paychecks. Not surprisingly, their analyses emphasize the positive and downplay the negative. The design process itself undermines the effects of social analysis because there is often an implicit analytical hierarchy in which technical analyses are taken more seriously than social analyses.

Second, there is a need for a much broader unit of analysis, as well as a more comprehensive approach. According to one well-placed professional within USAID:

> Social analysis brings with it a particular view of development. It has a bias in favor of promoting social equity and long-term growth. This is a "good bias" but it is a point of view not universally shared by people who make decisions in AID and other development agencies. . . . Social analysis needs to rethink its focus in order to retain or grow in importance. For example, it could play a bigger role in policy dialogue. A social analysis as part of policy dialogue would be a very useful contribution and an expansion into a new role.

This dissatisfaction stems basically from one hard, cold fact: over the past decade, USAID's development priorities have changed, as have those of other bilateral and multilateral donor agencies, and left social analysis behind, marking time, while economic analysis, as opportunistic as ever, has broadened its scope but continued to neglect some of the key issues traditionally raised by social analysis. Donor focus now is at the program level on policy reform, structural adjustment, privatization, entrepreneurship, capital markets, environmental sustainability, global agreements, and similar themes. But all of these policy interventions affect people—for better or for worse—whether they are put first or whether they come last (Cernea, ed. 1991; Chambers 1983).

Third, critics called for increasing emphasis on decision making, with a focus on the dynamics of power and process. Closely related was the call for more institutional analysis in both the public and the private sector. Overall, there was a continual push for social analysis to adapt to the development priorities of the 1990s and to become more integrated with the other analyses undertaken for project and program design.

The fact that social analysis has not lived up to expectations can be partially explained by the context in which it is undertaken. Success or failure in rural development has little to do with the quality of the social analysis in particular, or design documents in general, both of which are basically advocacy documents written to obtain funding, not planning documents to facilitate implementation. Serious planning only begins once the document has been approved and funding is available. Furthermore, given the uncertainties of project implementation, it is perhaps presumptuous to believe that anyone can realistically predict what will transpire.

The Proper Role: Broker, Collaborator, or Advocate?[6]

What, then, is the proper role of the anthropologist once the project is underway—participant employee or research evaluator? An analysis of the role and contribution of anthropologists to the planning and implementation of three rural development projects indicates the potentials and pitfalls of such close involvement. The projects were the Agroforestry Outreach Project (AOP) in Haiti, the Manantali Resettlement Project in Mali, and Project North Shaba (PNS) in Zaire, a large, integrated rural development project. All three involved the environment and natural resource management, and the provision of anthropological knowledge and understanding. I selected them because I knew them through personal involvement or personal interest; because anthropologists were believed to have made a significant contribution; and, finally, because there was good documentation available.[7]

Anthropologists have traditionally played the role of broker, translating the products of professional anthropology into normal discourse and bringing together groups who might not otherwise communicate. The information conveyed consists primarily of two components. First, it calls for understanding the ethnography of a particular project, particularly the relations between the project and the local and national political structures. Second, the anthropologist can pinpoint the likely consequences of certain actions and critically examine the assumptions on which a project is based by comparing it with similar cases in the development literature (Conlin 1985).

In the cases of Haiti and Zaire, project design and implementation were informed by prior anthropological knowledge of the area in question and, in the case of Mali, of similar resettlement experiences elsewhere. In all three, the anthropologists criticized prevailing models of development and suggested ways to improve project design and implementation.

Nevertheless, to be effective, anthropologists must be prepared to play a variety of roles, some of which are more compatible with professional ethics than others. Anthropologists may find themselves active participants in third world situations about which they may have profound reservations and misgivings. This is particularly the case with work done under contract, where keeping the customers happy always has a high priority so that they will request more of the same (Gow 1991). You try to avoid biting the hand that feeds you while preserving some vestiges of independence and professional integrity.

Participation leads to collaboration, which begets compromise, which ultimately may lead to co-option. In all three cases, as the anthropological involvement was funded by USAID, there was collaboration of various degrees. But the more active the involvement of the anthropologist, the more strained the collaboration, partly because the anthropologists felt they were being asked to do more than they thought they legitimately should. In the case of Haiti, this collaboration was tempered by the fact that the whole undertaking bypassed the national government, a predatory institution strongly condemned by Haitians and expatriates alike. Anthropologists who knew the country well were given positions of responsibility within the decision-making structure of the project because they had the language and the experience to function effectively in the field. The organizational structure, together with the wide geographical dispersion of project activities, permitted them a great deal of freedom and flexibility. As the project became more institutionalized and the need was more for bureaucrats and managers, the anthropologists were phased out.

In contrast, the situations in Mali and Zaire were more tightly structured. USAID was working directly with the respective national governments, marching to an agenda established in Washington, DC. In both cases, the most important contribution was provision of anthropological knowledge, information that significantly affected the way the activities were designed and implemented. In Mali, this meant that the project was designed to address a social, as opposed to an engineering, issue—how to resettle the local population without making it totally dependent on the state. During the resettlement, the anthropologists were there to monitor implementation, collect information on specific activities, and present it to decision makers.

In Zaire, a similar transformation occurred during the design phase. Instead of the traditional authoritarian model favored by the government, USAID designed a more participatory project that respected existing forms of agricultural production and social organization. During implementation, the anthropologists were directly involved, but as managers, with little prior knowledge or understanding of the country. In this sense, they were operating as "hired guns," albeit from an anthropological perspective (Gow 1991). On the other hand, given the isolation of the project, they and their Zairian colleagues did enjoy considerable autonomy in making decisions and allocating resources at the project level.

Collaboration was, of course, tempered by the advocacy role played by anthropologists in all three projects—defending the rights of the local population in Haiti and Mali, more properly defending the interests of the project per se and its employees in Zaire. Though advocacy within the field of anthropology continues to receive a mixed press, the indications are that this is a role relished primarily by those who have made a long-term personal and professional commitment to the country and people in question (Gow 1993).

Moving with the Times: The Anthropologist as Policy Maker

Some believe, like the USAID official quoted earlier, that anthropology should be capable of formulating policies that will contribute to the understanding of contemporary development issues, particularly the social, political, environmental, and institutional dimensions (Weaver 1985). In all three cases, the acid test of anthropological effectiveness was in the policy domain. Here, leverage was strongest and most evident in the design phase of all three projects. In Haiti, USAID policy was strongly influenced in choice of strategy—trees as crops—and choice of institutions—nongovernmental organizations (NGOs) rather than public entities. In Mali, settlement policy was broadened to include social, tenurial, and dependency aspects, supported by an impressive body of long-term, comparative anthropological research.[8] In Zaire, project policy incorporated agroecological variation, existing farming systems, and a healthy respect for the recent historical past. During implementation, effectiveness was somewhat curtailed, particularly in Mali and Zaire, where, in spite of reasoned arguments to the contrary, USAID insisted on following guidelines dictated by Washington. In Haiti, in contrast, anthropologists were still able to significantly affect policy, particularly in establishing priorities and strategies for the second phase of the project.

Part of the ability to influence policy for the better was due to the authority of the anthropologists: they had been there, they knew the people and the terrain, they spoke on their behalf. In other words, they were credible. The problems came when higher authorities set the agenda without considering local conditions or realities, as in Mali and Zaire. Ultimately, anthropologists were unable to change those policies. Their anthropological authority paled beside that of the political authority, whether in Washington, Bamako, or Kinshasa. Weaver (1985) has argued that some of the difficulty anthropologists have experienced in effectively addressing these issues has its roots in a lack of familiarity with how policy is made, as well as inattention to the bureaucratic and administrative contexts in which policy is formulated. Although this is a valid criticism, the issue here is not lack of knowledge, but lack of power.

Though anthropologists may have made important contributions to the study of power, professionally they have chosen, with several notable exceptions, to occupy a marginal position, whether within the power structure of the nation or within the academic community (Hoben 1984). For those in academia, at the risk of oversimplification, power accrues from the institution where one works, the grants and contracts one receives, the journals and presses where one publishes, and the conferences to which one is invited. And the basis and security of this power is tenure.

For practicing anthropologists, the situation is markedly different. Permanent employment in one of the development bureaucracies, either public or private, may be problematic as one is transformed from an anthropologist into an administrator. For administrators, the political and pragmatic interests of their own organizations, particularly their own positions within those organizations, are major constraints on action. Thus, the institutional context will often determine and, to a certain extent, dictate the limits of both thought and action:

> The ideal administrator is intelligent, knowledgeable, balanced, consistent, predictable, reliable, moderate in speech, and a team player who seeks to follow precedent and find common grounds for action in the differing beliefs and interests of those with whom he or she works. [Hoben 1984:11]

In other words, the institution, rather than the individual, sets the agenda.

There have been repeated appeals over the years for anthropologists to cut their hair, put on a suit, and join the larger institutions working in development, both public and private (Cernea 1991). And many have done this—with varying degrees of success and effectiveness. But with the exception of those who manage to obtain a research position, the great danger is that one will lose one's anthropological perspective.[9] Working as a contractor, as a "participant employee," though an invaluable ethnographic experience that can contribute greatly to increasing the anthropologist's credibility, changes little in the long run. And contractors, by definition, have no tenure.

For those without some permanent institutional affiliation, such power as they enjoy is often ephemeral, a combination of independence from many of the normal bureaucratic controls, recognition as an "expert in the field," and what I can only term the challenge and excitement of what is often a total experience. But without some independent means of support, such a way of life has severe limitations. Full-time consulting is a very uncertain business that places great demands—both professional and psychological—on the individual involved (Frank 1987; Robins 1986).

Principles, Power, and Praxis

When all is said and done, the environment is still about people. In spite of the quotations at the beginning of this article, we are not witnessing the end of nature—at least not yet. What we are seeing is a growing awareness of the relationships between people and their environment, linkages that require the perspectives of the social scientists and the natural scientists, as well as agreement on some principles that can guide practical solutions. I have already argued that there is a growing consensus within the environmental community about the importance of people, in cultural, political, and social terms—the traditional domain of the anthropologist. To a certain extent, anthropologists have risen to the challenge, but it is natural science and politics, not social science, that is setting much of the environmental agenda for the 1990s.

Anthropology can continue with business as usual or begin to address some of the issue raised here, building on some of the more stimulating changes presently underway in the profession. Until recently, much of the writing in anthropology on third world issues has been for fellow academics or for the customer, the institution that finances the work, with a little for academic consumption, rather than for the subject, the people presented and discussed in the report:

> The ability to perceive "from the native's point of view" . . . not only assumes that the natives need someone to "speak for" them, but also becomes entangled with the need to perceive "from the institution's point of view." Moreover, the anthropologist thus constructs local situations not in terms that conform to the subject's reality but in terms that make sense to the organization, that is, a reflection of development categories. [Escobar 1991:672–673]

Anthropologists, however, are, albeit slowly and perhaps reluctantly, beginning to write for the people they study, attempting to meet their needs for information and suggesting ways in which they can effectively use this information (Rappaport 1994). In addition, as the Earth Summit demonstrated, third world people are making their voices heard in broader arenas through various forms of local organizations, often in association with anthropologists. Over the past two decades, international and national NGOs have come to play an increasingly prominent role in environmental and developmental issues, as advocates, intermediaries for local people, and implementers of projects. Partly through choice and partly through necessity, anthropologists are becoming involved with NGOs. NGOs, however, should not be viewed as some panacea for all environmental problems in the third world (Farrington et. al 1993). NGOs and anthropologists share many of the same strengths and weaknesses. To be effective, such involvement must be more than the conventional focus on "bottom-up" development and include the political and institutional constraints that prevent local people from realizing their potential, the policy arena that engaged the anthropologists in the three case studies. Closely related is the increasing concern with local people's access, use, and control of renewable resources—in many senses the basis of *their* power (Bowen 1988). In practice, this means that anthropologists should concentrate on assisting local people to develop their productive resources. Where resources are limited or insufficient, anthropologists should help create new assets, in this way providing a more favorable context for improved stewardship of the natural resource base and strengthening local people's livelihood security. Thus the practice of participation is broadened to include empowerment, with both economic and political implications, the goal of which—over the long run—is to change society.

Political economy can contribute to an understanding of such problems, with its underlying premise that politics and economics, polity and economy, are inextricably linked and must be examined as a whole. Political ecology combines the concerns of ecology and a broadly defined political economy, particularly a concern with the role of the state, which

> commonly tends to lend its power to dominant groups and classes, and thus may reinforce the tendency for accumulation by these dominant groups and marginalization of the losers, through such actions as taxation, food policy, land tenure policy and the allocation of resources. [Blaikie and Brookfield 1987:17–18]

Ecology alone is insufficient for understanding many kinds of natural resource management problems because it often neglects politics, power, ethics, sociocultural values, and conflicts that influence the use of resources (Stonich 1992). In practice, this means identifying the various groups and interests involved in the development process, assessing their degree of influence and control of resources, and tracing the linkages among them. In other words, politics must be taken seriously and treated with respect, in contrast to the prevailing approach, where politics is often accepted as a given.

In sum, if practicing anthropology is to make a meaningful contribution to the analysis and solution of environmental problems, both national and international, then it must combine theory and practice, acknowledge the importance of certain guiding principles, and confront the issue of power—both professional and personal—head on. An analytical framework that draws on political ecology can help, as can a more critical awareness of both the substance and process of sustainable development in the third world. Finally, a large dose of humility is in order. The age of "international experts" is passing—they are an endangered species whose days are rightfully numbered. Solving environmental issues, and developmental problems in general, calls for closer collaboration, greater respect, and more willingness to share information and expertise between the North and the South, between donors and national governments, and, most important—and perhaps more realistically—among fellow anthropologists and the institutions and people they wish to strengthen.

Notes

Acknowledgments. An earlier version of this article was presented at the annual meeting of the Society for Applied Anthropology held in Memphis in March 1992. This version draws heavily on my own recent writings on anthropology, natural resource management, and the environment. See particularly Gow 1992a, 1992b, and 1993.

1. This is a vast topic, with a burgeoning literature. The first perspective, which is currently more common in both the literature and the profession, is represented by Cernea 1991; the second perspective is exemplified by Escobar 1991.

2. But this is steadily changing. In the summer of 1992 I was working in Anglophone Africa. During my stay in Uganda, where I was working on the National Environmental Action Plan, I noticed articles—in both the government press and the opposition—strongly criticizing the nature and substance of international assistance, starting with the World Bank and the

International Monetary Fund, and working down to the large salaries paid to so-called expatriate experts. The bases of the comments were usually three: first, this assistance has produced little direct benefit for the majority of Ugandans; second, the policies and priorities were often established by the donors, not by the national government; third, the money spent on technical assistance could be much more profitably spent employing qualified Ugandans or invested in more directly productive activities.

3. This is discussed in more detail in Gow 1992a.

4. This reinforces an important distinction between the North and the South. In the third world, generally speaking, the concept of the environment is much broader than in the North and includes not only the conventional environmental concerns, from pollution to biodiversity, but also those of natural resource management in general, including deforestation, soil erosion, overgrazing, encroachment, and poaching.

5. These ideas are presented in more detail in Gow 1992b.

6. This threefold division was inspired by the work of Whisson (1985).

7. For Haiti, see Lowenthal 1990 and Murray 1979, 1987; for Mali, see Koenig 1986 and Koenig and Horowitz 1990; and for Zaire, see Barclay 1979, Blakely 1982, and DeLucco 1988. I discuss and analyze these projects in more detail in Gow 1993. The analysis that follows draws heavily on the latter.

8. The Institute for Developmental Anthropology (IDA) has been very active in drawing attention to the problems encountered in forced resettlement projects, particularly the work of Scudder (1985). The World Bank has also been active in this domain and has issued guidelines for assessing and mitigating the impacts of what it terms "involuntary resettlement" (Cernea 1988).

9. The literature is ominously quiet about what it is really like to be an anthropologist working as a permanent staff member within a large bureaucracy. The articles available have usually been written by anthropologists who are still functioning members of the bureaucracy. Thus, the necessary objectivity and critical self-awareness, together with insider insights, may be lacking. For example, the best accounts I know of how USAID functions, Hoben 1980, 1989 and Tendler 1975, were written once the authors were no longer employees of the institution.

References Cited

Barclay, A. H.
    1979 Anthropological Contributions to the North Shaba Rural Development Project. Paper presented at the annual meeting of the Society for Applied Anthropology, Philadelphia. Washington, DC: Development Alternatives, Inc.
Bennett, J. W.
    1988 Anthropology and Development: The Ambiguous Engagement. In Production and Autonomy: Anthropological Studies and Critiques of Development. J. W. Bennett and J. R. Bowen, eds. Pp. 1–29. Monographs in Economic Anthropology, 5. Lanham, MD: University Press of America.
Blaikie, P., and H. Brookfield, eds.
    1987 Land Degradation and Society. New York: Methuen.
Blakely, T. D.
    1982 Achievements and Challenges in Small Farmer Agricultural Development: The North Shaba Project Area in 1982. Report prepared for AID/Zaire. Kinshasa: AID.
Bowen, J. R.
    1988 Power and Meaning in Economic Change: What Does Anthropology Learn from Development Studies? In Production and Autonomy: Anthropological Studies and Critiques of Development. J. W. Bennett and J. R. Bowen, eds. Pp. 411–430. Monographs in Economic Anthropology, 5. Lanham, MD: University Press of America.
Burdge, R. J.
    1990 The Benefits of Social Impact Assessment in Third World Development. Environmental Impact Assessment Review 10:123–134.
Cernea, M. M.
    1988 Involuntary Resettlement in Development Projects. World Bank Technical Paper 80. Washington, DC: World Bank.

1991   Knowledge from Social Science to Development Policies and Projects. *In* Putting People First: Sociobiological Variables in Rural Development. 2nd ed. M. M. Cernea, ed. Pp. 1–41. A World Bank Publication. New York: Oxford University Press.

Cernea, M. M., ed.
1991   Putting People First: Sociological Variables in Rural Development. 2nd ed. A World Bank Publication. New York: Oxford University Press.

Chambers, R.
1983   Rural Development: Putting the Last First. New York: Longman.

Colby, M. E.
1990   Environmental Management in Development: The Evolution of Paradigms. World Bank Discussion Paper 80. Washington, DC: World Bank.

Conlin, S.
1985   Anthropological Advice in a Government Context. *In* Social Anthropology and Development Policy. R. Grillo and A. Rew, eds. Pp. 73–87. ASA Monographs, 23. London: Tavistock Publications.

DeLucco, P.
1988   The End of the Road: The North Shaba Integrated Rural Development Project Approaches PACD. Paper prepared for the John F. Kennedy School of Government, Harvard University.

Escobar, A.
1991   Anthropology and the Development Encounter: The Making and Marketing of Development Anthropology. American Ethnologist 18(4):658–682.

Farrington, J., and A. Bebbington, with K. Wellard and D. J. Lewis
1993   Reluctant Partners? Non-Governmental Organizations, the State, and Sustainable Agricultural Development New York: Routledge.

Forché, C.
1981   The Country between Us. New York: Harper and Row.

Frank, L.
1987   The Development Game. Granta (April):229–243.

Gardner,J.E.
1989   Decision Making for Sustainable Development: Selected Approaches to Environmental Assessment and Management. Environmental Impact Assessment Review 9:337–366.

Gow, D. D.
1991   Collaboration in Development Consulting: Stooges, Hired Guns, or Musketeers? Human Organization 50(1):1–15.
1992a Poverty and Natural Resources: Principles for Environmental Management and Sustainable Development. Environmental Impact Assessment Review 12(1/2):49–65.
1992b Forestry for Sustainable Development: The Social Dimension. Unasylva 43(2):41–45.
1993   Doubly Damned: Dealing with Power and Praxis in Development Anthropology. Human Organization 52(4):380–397.

Gow, D., C. Haugen, A. Hoben, M. Painter, J. VanSant, and B. Wyckoff-Baird
1989   Social Analysis for Third World Development: Toward Guidelines for the Nineties. Report prepared for AID/Washington. Washington, DC: Development Alternatives/Institute for Development Anthropology.

Hoben, A.
1980   Agricultural Decision Making in Foreign Assistance: An Anthropological Perspective. *In* Agricultural Decision Making: Anthropological Contributions to Rural Development. P. F. Barlett, ed. Pp. 337–369. New York: Academic Press.
1984   Role of the Anthropologist in Development Work: An Overview. *In* Training Manual in Development Anthropology. W. L. Partridge, ed. Pp.9–17. Washington, DC: American Anthropological Association.
1989 USAID: Organizational and Institutional Issues and Effectiveness. *In* Cooperation for International Development: The United States and the Third World in the 1990s. R. J. Berg and D. F. Gordon, eds. Pp. 253–278. Boulder, CO: Lynne Rienner Publishers.

International Union for the Conservation of Nature and Natural Resources
1980   World Conservation Strategy. Gland, Switzerland: IUCN.

International Union for the Conservation of Nature and Natural Resources, United Nations
Environmental Program, and World Wildlife Fund
    1990   Caring for the Earth: A Strategy for Sustainable Living. Summary. Gland, Switzer-
       land: IUCN/UNEP/WWF.
Jacobs, P., J. Gardner, and D. A. Munro
    1987   Sustainable and Equitable Development: An Emerging Paradigm. In Conservation
       with Equity: Strategies for Sustainable Development. P. Jacobs and D. A. Munro, eds.
       Pp. 17–29. Cambridge/Gland: Cambridge University Press/IUCN.
Koenig, D.
    1986   Research for Rural Development: Experiences of an Anthropologist in Rural Mali.
       In Anthropology and Rural Development in West Africa. M. M Horowitz and T. M. Painter,
       eds. Pp. 29–60. Boulder, CO: Westview Press.
Koenig, D., and M. M. Horowitz
    1990   Involuntary Settlement at Manantali, Mali. In Social Change and Applied Anthropol-
       ogy: Essays in Honor of David W. Brokensha. M. S. Chaiken and A. K. Fleuret, eds. Pp.
       69–83. Boulder, CO: Westview Press.
Kottak, C. P.
    1991   When People Don't Come First: Some Sociological Lessons from Completed
       Projects. In Putting People First: Sociological Variables in Rural Development. 2nd ed.
       M. M. Cernea, ed. Pp. 431–465. A World Bank Publication. New York: Oxford University
       Press.
Lele, S. M.
    1991   Sustainable Development: A Critical Review. World Development 19(6):607–621.
Little, P., ed.
    1991   Development Anthropology Network 9(1). (Special number devoted to the first 15
       years of the Institute for Development Anthropology.)
Lowenthal, I.
    1990   National Program for Agroforestry in Haiti: Social Soundness Analysis. In National
       Program for Agroforestry in Haiti, vol. 2. DESFIL, ed. Pp. 1–47. Washington, DC:
       Development Alternatives.
McKibben, W.
    1990   The End of Nature. Harmondsworth, UK: Penguin Books.
Morgan, E. P.
    1985 Social Analysis and the Dynamics of Advocacy in Development Assistance. In Social
       Impact Analysis and Development Planning in the Third World. W. Derman and S.
       Whiteford, eds. Pp. 21–31. Boulder, CO: Westview Press.
Murray, G. M.
    1979   Terraces, Trees, and the Haitian Peasant: An Assessment of 25 Years of Erosion
       Control in Rural Haiti. Report prepared for AID/Haiti. Port-au-Prince: AID.
    1987   The Domestication of Wood in Haiti: A Case Study in Applied Evolution. In Anthro-
       pological Praxis: Translating Knowledge into Action, R. M. Wulff and S. J. Fiske, eds. Pp.
       223–240. Boulder, CO: Westview Press.
Myers, N.
    1989   The Environmental Basis of Sustainable Development. In Environmental Manage-
       ment and Economic Development. G. Schramm and J. R. Warford, eds. Pp. 57–68.
       Published for the World Bank. Baltimore: The John Hopkins University Press.
Rappaport, J.
    1994   Cumbe Reborn: An Andean Ethnography of History. Chicago: University of Chicago
       Press.
Robins, E.
    1986   Problems and Perspectives in Development Anthropology: The Short-Term Assign-
       ment. In Practicing Development Anthropology. E. C. Green, ed. Pp. 67–76. Boulder,
       CO: Westview Press.
Scudder, T.
    1985   A Sociological Framework for the Analysis of New Land Settlements. In Putting
       People First: Sociological Variables in Rural Development. M. M. Cernea, ed. Pp.
       121–153. New York: Oxford University Press.

Stonich, S. C.
    1992 Struggling with Honduran Poverty: The Environmental Consequences of Natural
        Resource-Based Development and Rural Transformations. World Development
        20(3):385–399.
Tendler, J.
    1975 Inside Foreign Aid. Baltimore, MD: Johns Hopkins University Press.
Weaver, T.
    1985 Anthropology as a Policy Science: Part 1, A Critique. Human Organization 44(1):97–
        105.
Whisson, M. G.
    1985 Advocates, Brokers, and Collaborators: Anthropologists in the Real World. *In* Social
        Anthropology and Development Policy. R. Grillo and A. Rew, eds. Pp. 131–147. ASA
        Monographs, 23. London: Tavistock Publications.
World Resources Institute, International Union for the Conservation of Nature, and United
Nations Environmental Program
    1992 Global Biodiversity Strategy: Guidelines for Action to Save, Study, and Use Earth's
        Biotic Wealth Sustainably and Equitably. Washington, DC: WRI.
World Commission on Environment and Development
    1987 Our Common Future. Oxford, UK: Oxford University Press.
Wright, R. M.
    1991 Introduction. *In* Wildlands and Human Needs: Reports from the Field. R. D.Stone,
        ed. Pp. 1–4. Washington, DC: World Wildlife Fund.

The Tropical Forestry Action Plan: Is It Working?

Robert Winterbottom

Introduction

The Tropical Forestry Action Plan (TFAP) was formulated as an international framework for the development of country-level forestry action plans in response to the crisis of tropical deforestation (Food and Agriculture Organization 1985; World Resources Institute 1985). These national plans aim to control tropical deforestation and promote the economic benefits of forestry as a result of increased development assistance and support for activities in five critically important areas: forestry in land use, especially agricultural land use; fuelwood and wood energy; forest-based industrial development; conservation of forest ecosystems; and institution strengthening, especially in the areas of research, training, and extension. The process of preparing and implementing such national plans raises many of the constraints and challenges inherent in promoting "sustainable development."

A key stumbling block for the TFAP process has been a preoccupation with investment-oriented technical solutions identified by specialists within the forestry sector. People, instead of policies and politics, were seen as the primary cause of deforestation and mismanagement of forests. The insufficient involvement of anthropologists and other social scientists, as well as a lack of effective mechanisms for local participation, resulted in an inadequate analysis of many factors that exert a powerful influence on patterns of forest exploitation.

The TFAP is, nonetheless, an example of concerted action by national governments and development assistance agencies to address the continued loss and mismanagement of tropical forests. In 1991, over 65 countries were engaged in the TFAP planning process, and at least one billion dollars per year was being mobilized for development assistance in forestry. The basic principles and guidelines for country-level TFAP planning exercises have been outlined and revised over the past five years. In a number of countries, special roundtable meetings and workshops have been organized to increase participation in national TFAP exercises. However, the tendency to implement the TFAP as an extension of development assistance in the forestry sector has favored control and management by the forestry profession and forestry agencies. The control of the planning process by foresters has, in turn, severely constrained the ability of the TFAP to examine the need for institutional reforms within the forestry agencies, or to analyze linkages between sectors, promote broad participation, and identify needed policy reforms both inside and outside the forestry sector.

Of the many suggestions for the improvement of the TFAP made over the past year, the following are highlighted in this article:

- Increased support for multidisciplinary analysis, multisectoral approaches, and broad-based participation in TFAP exercises, including improved mechanisms for the involvement of anthropologists and other social scientists.

- Greater attention to quality control in TFAP planning exercises.
- More emphasis on policy reforms and development strategies and less attention in the early stages of the planning process to the technical details of project identification and investment plans.
- Special attention and mechanisms to address the concerns and needs of indigenous peoples, forest-dwellers, the rural poor, and other marginalized groups that generally are not well-represented in development planning.

The TFAP planning framework recognizes, in principle, that the fundamental causes and driving forces of deforestation are enmeshed in a web of complex social, political, and economic issues. The process of preparing a national TFAP was designed to examine these interrelated issues or "root causes," identify appropriate development strategies, policy reforms, and investment needs, and promote a concerted effort to mobilize the financial resources and other actions needed to provide for the sustainable development of a country's forest resources. At the same time, the implementation of national TFAPs was intended to contribute to improved food security and rural land use, increased supplies of forest products, and increased income and well-being for people dependent on forest resources (FAO et al. 1987).

As this brief analysis makes clear, the experience to date with the TFAP has revealed that the role of social scientists (and anthropologists in particular) must be strengthened in the future in the TFAP planning process. Their contribution is essential for the simple reason that "solutions" to the deforestation crisis lie for the most part outside the realm of expertise of technical specialists in forestry. Development policies, politics, and people are the key areas that need to be thoroughly examined if we wish to have any success in controlling deforestation. Before the current patterns of forest resource exploitation can be changed, the expertise of anthropologists, economists, and many other social scientists is needed to more fully understand how the operative incentives at all levels can be modified to favor more sustainable patterns of resource use and management.

Unfortunately, in many early TFAP exercises, the need for policy reform was neglected, and people were seen as the source of the problem—the agents of forest destruction. This perspective neglected to recognize government policies and other socioeconomic causes that exacerbated the problems of forest encroachment and over-harvesting of forest resources. These causes include skewed land ownership, insecure land tenure, the need to clear forest to obtain land title, and tax breaks or other incentives for forest clearance and wasteful production of timber (Mahar 1989; Repetto 1990; Repetto and Gillis 1988). Furthermore, the initial approach used to prepare TFAPs neglected an objective reassessment of the impacts of forest-dwelling people, traditional shifting cultivators, and other local communities that had, in fact, played an important role in the sustainable, rational use of forest lands and forest resources (Colchester and Lohmann 1990; Lynch 1990). The particular needs and rights of indigenous peoples, in particular, have been virtually overlooked in many national TFAPs (Cort 1991: Halpin 1990).

The Deforestation Crisis

Over half of the world's remaining forests are found in tropical, developing countries. Their rapid conversion and continuing degradation has created a crisis of global proportions. Fewer than one billion hectares remain of the approximately

1.6 billion hectares of moist tropical forest that once existed. Although the data are still imprecise in many countries, the World Resources Institute (WRI) estimates that the remaining tropical forests are being cleared at a rate that exceeds 20 million hectares annually (WRI 1990a). This rate of deforestation is equivalent to clearing the entire forest area of Gabon every year, or about 40 hectares a minute.

In many cases, the clearing and conversion of tropical forests does not lead to a sustained increase in agricultural productivity. Unlike the case of forest conversion in temperate climate, the conversion of tropical forests and clearance of tropical soils to sustainable, alternate forms of land use is constrained by a number of ecological factors that—unless adequately addressed—result in a decline in the long-term productivity of the land resource.

Forest loss is particularly rapid and worrisome in Central America, in the Amazon basin, in Thailand, Malaysia, Madagascar, and in several west African countries (Nigeria, Cote d'Ivoire, Liberia). An unknown, and significantly larger, area is being steadily degraded through shortened periods of forest fallow and intensified shifting cultivation, uncontrolled burning, overgrazing, nonsustainable fuelwood harvesting, drought, and other pressures.

Less than 1 percent of the productive moist tropical forests are being operationally managed for the sustained yield production of industrial wood; this amounts to fewer than one million hectares out of 828 million hectares of commercially productive forest land (Poore et al. 1989). Reforestation and forest regeneration amount to less than one-tenth of the area annually deforested (FAO 1982).

The Human Costs and Socioeconomic Implications of Deforestation

At the local level, the decline in the extent and productivity of the forest resource base is often linked to reduced food security; increased vulnerability to drought, flooding, and soil erosion; disruption of supplies of food and water, fodder, medicinal plants, and raw materials for cottage industries and local market produce; reduction in incomes and employment opportunities; and a host of other perturbations. The very existence or future livelihoods of forest-dwelling tribal groups is also often threatened (Cabarle 1989; Halpin 1990).

At the national level, foreign exchange earnings from timber exports can be jeopardized, and growing scarcities of wood fuels often lead to increased dependence on imported commercial fuel substitutes and fossil fuels. Expensive investments in transportation, power generation, and irrigated agriculture are likely to be adversely affected as dams silt up more rapidly and reservoir storage capacity is diminished by deforestation in upstream catchment areas. The aggregate impact of deforestation on rural economies and industrial employment can be severe. At the global level, significant and irreplaceable biological resources are lost, genetic diversity declines, and the problems of global warming and sea level rise are exacerbated (Myers 1989; Reid and Miller 1989).

The International Response: The TFAP

In an effort to move beyond an assessment of the problem of tropical deforestation, the WRI, in cooperation with the World Bank and the United Nations Development Programme (UNDP), convened an international task force in December 1984 to examine possible responses to the tropical deforestation crisis. Within six months, this task force had agreed upon a broad framework for action (WRI 1985).

At the same time, the Forestry Department of the Food and Agriculture Organization (FAO) was developing a tropical forestry action program under the auspices of the Committee on Forest Development in the Tropics (FAO 1985).

Both plans were presented and discussed at the World Forestry Congress in June 1985 in Mexico. By November the concept of TFAP had been endorsed by most major aid agencies and numerous national governments. A summary version of the TFAP prepared by WRI, FAO, the World Bank, and UNDP was endorsed in 1987 by a high-level conference of political decision makers and has subsequently been endorsed by the heads of state summit conference in Paris and other high-level conferences and meetings (Bellagio Meeting 1987).

One of the stated goals of the TFAP is "to reverse the process of deforestation" and to "harness" the potential of the forest resources of developing countries to meet their development needs (FAO et al. 1987). It aims to conserve tropical forests as an "essential resource for the economic and social well-being of rural people in developing countries" by overcoming the "lack of political, financial and institutional support to apply solutions."

The TFAP framework document especially urges avoidance of "costly mistakes associated with past emphasis on massive development projects" and highlights the need for better planning of projects outside the forestry sector that might otherwise jeopardize tropical forests. The plan points to the need for practical strategies and approaches to reforestation and forest management, which involve the "millions of people who live within and beside the forests and depend upon them to help satisfy their basic needs", and which take advantage of the "important role" that nongovernmental organizations (NGOs), grassroots organizations, and local communities need to play. The TFAP also emphasizes that "action is needed now" in addressing the "cycle of destruction" of tropical forests.

The general description of the TFAP framework also draws attention to a number of underlying causes of deforestation, including rapid population growth. It points to the direct impact of shifting agriculture, conversion to permanent agriculture, over-exploitation of fuelwood stocks, and careless logging of the remaining tropical forests. It also notes the likely consequences of a failure to address these fundamental causes in terms of depleted fuelwood supplies, disruption of indigenous forest dwellers, loss of biodiversity, degradation of watershed and agricultural soils, and destruction of timber and related economic benefits.

Implementation of the TFAP

Beginning in 1985, an informal group of representatives from the major bilateral and multilateral aid agencies began to meet at six-month intervals to coordinate and support the implementation of the TFAP at the national level. A special coordinating unit for the TFAP was established within the Forestry Department of the FAO, and national steering committees were organized in a number of countries to guide the preparation and implementation of national TFAPs.

Guidelines for the development of national TFAPs have been drafted by FAO, revised according to the experience gained in the first few years of preparing TFAPs, and expanded to take account of the need for broad and comprehensive analysis of the underlying factors related to deforestation and inadequate management of forest resources (FAO Forestry Department 1989a). The FAO guidelines propose an 18-month planning process, beginning with the decision of the government to develop a TFAP, followed by the preparation of a paper highlighting the main issues to be addressed. Detailed terms of reference are then developed for the preparation of a forestry sector review. This review is formally presented to a roundtable meeting of interested government agencies and donors, together with an investment plan and proposed priority projects.

By September 1987, some 30 countries had begun the process of preparing national TFAPs; by early 1990, more than 65 countries had initiated or completed the preparation of national TFAPs. Among the 74 countries that have initiated national forestry sector reviews and other planning activities related to the preparation of national TFAPs, 34 have completed sector reviews, and 24 have presented their completed action plans to development assistance agencies (FAO Forestry Department 1991).

Basic Principles of the TFAP

By 1989, the FAO and the TFAP forestry advisors decided to reexamine the objectives of the TFAP framework as a planning process at the national level. This was prompted both by a need to clarify the strategic orientation of the TFAP and by the need to identify more specific targets against which progress could be measured. According to FAO, the primary objectives are still "rural development (food security, alleviation of poverty, equity and self-reliance) and sustainability of development (ecological harmony, renewability of resources, conservation of genetic resources)." However, a series of "basic principles" have also been outlined "which characterize the TFAP strategy in reaching its ultimate objective of conservation and development of tropical forest resources" (FAO Forestry Department 1989b). These principles are:

- Declared political commitment at higher government level to forestry and its role in rural development.
- Forestry policies that reconcile the present and future needs for sustainable development and the environmental role of forests and trees, and that focus on meeting the needs of local people, particularly the rural poor, who depend on forest and tree resources for their subsistence and food security and as a major opportunity for development.
- A visible role for forestry in national development plans with a clear indication of objectives, priorities, and allocation of increased public resources.
- Active, organized, and self-governed involvement of local groups and communities in forestry activities, with a particular focus on the most vulnerable, on women, and on commonly shared resources.
- The involvement of rural people, local NGOs, and the private sector in the planning and management of forestry activities.
- An identification of the main constraints and problem areas requiring immediate action to restore or maintain the critical role of forestry in environmental stability and socioeconomic development.
- A systematic combination of actions geared at monitoring and conserving the resource base and at raising and broadening the goods and services produced by forests and trees.
- Effective coordination of policy, planning, and implementation of activities among relevant departments involved in the primary use of land such as agriculture, livestock, forestry, mining, energy, and so on, and also with those involved in processing (cottage and industry) and commerce.

- Increasing public, private, national, and international investments to increase the production of goods and services from forestry.
- An effective and increased support by the international community based on a concerted response to technical and financial assistance needs and priorities expressed by tropical countries in line with the principles of TFAP.

Despite repeated statements of principles and the existence of carefully crafted guidelines, the extent to which the TFAP has actually followed the principles and the congruence of national strategies with the proposed objectives, strategic framework, and guidelines outlined by FAO has not, however, been comprehensively or systematically monitored by the TFAP coordinators (Winterbottom 1990).

Preliminary Analysis of Major Issues Addressed by the TFAP

As I have described, the TFAP was an attempt to respond more effectively to the deforestation crisis, primarily by increasing the impact of forestry development aid. Such aid was to be linked with developing more comprehensive programs for improved forest resources management and integrating these programs into broader strategies for sustainable development at the national and international levels. Local participation, national leadership, and country-level ownership of the process were all seen as essential to planning and implementing the TFAP.

However, a recurrent source of trouble for the TFAP has been its emphasis on increased forestry sector investment and control of the TFAP planning process by the FAO Forestry Department and chief forestry advisors of the aid agencies. It has become clear to all involved in the TFAP planning process that in order to succeed, the TFAP must address what are essentially economic and political problems which extend beyond the limited context of development assistance in the forestry sector.

How can forestlands be better used to meet the increasing needs for food, fuel, and other economic imperatives of developing countries without sacrificing the diversity and productivity of the forests? Can forestry development be compatible with the maintenance of livelihoods of traditional forest-dwelling people? Finding answers to these questions requires the direct involvement of a wide range of institutions and sectors, including agriculture, water resources development, transportation, infrastructure, population and health, industries, and social affairs. One cannot expect that foresters alone can solve the underlying problems driving deforestation, even if they manage to double investment in forestry (from its current low level of only 2 percent of the total budget for development assistance). More important, if the TFAP planning process cannot influence what happens with the other 95 percent of development assistance and what happens outside of the development assistance process, there will be little prospect of a significant, long-term impact on the deforestation crisis.

Another manifestation of the TFAP's weakness in addressing the underlying sociopolitical issues related to deforestation has been its "technocratic" orientation to the problem. From the forester's perspective, deforestation was seen to be primarily the result of increased population pressure, encroachment of farmers on forest lands, and over-harvesting of forest resources. From their perspective, the preferred solution was to invest in planting more trees, improved management of larger areas of forest, expansion of schools to train foresters, and so on. Had local communities and other disciplines, such as anthropology, had a larger role in the TFAP process, the emphasis of the proposed "actions" in national TFAP would no doubt have been more attuned to the underlying causes of deforestation and mismanagement of forests. These include poverty, inequitable land distribution, low agricultural productivity, poor land use policies, and inappropriate or poorly designed development projects.

In retrospect, some of the major constraints or challenges that the TFAP has faced include:

- The need to probe deeply into the factors driving deforestation and to identify appropriate measures to address these factors, even if they run counter to the short-term interests of politically powerful elites in government and the private sector.
- A lack of effective measures to build political commitment among a broad coalition of decision makers, who may be preoccupied with other issues or even threatened by a close look at forestry issues.
- A lack of suitable techniques to resolve trade-offs between local and national interests, between the needs of today versus tomorrow, between production versus protection, between short-term economic gains and longer-term ecological considerations.
- The need for suitable mechanisms to increase participation in the planning and implementation of the TFAP from a wide range of public agencies, private sector interests, and "independent voices," notably indigenous peoples and displaced colonialists, with adequate representation provided for the landless, women, and other frequently disenfranchised groups.
- The need for more integrated strategies, incorporating actions inside and outside the forestry sector based on analysis from a range of disciplines.
- The need to accelerate information collection and analysis, human resources development, and transfer of technologies while providing for sustained institutional growth and increased capacity at the national and local levels.
- The need to stimulate actions across a broad front to directly and indirectly address the root causes of deforestation and wasteful use of forest resources; actions are needed that go beyond a review of investment needs to encompass policy reforms, institutional restructuring and coordination, and incentives for sound land use and sustainable management of forests.
- The need to overcome institutional inertia and bureaucratic delays that impede the transfer of resources that are required, frustrate attempts to change the quality of aid, and stall the development assistance planning process.

Clearly, these issues cannot be adequately resolved by a small group of forestry professionals operating under the supervision of a national forestry department.

Accomplishments of the TFAP

Despite many difficulties and shortcomings in its implementation, the TFAP did create a forum for regular consultation among the principal development assistance agencies, and promoted collaboration among these agencies on forestry sector programs. The TFAP has helped to increase the number of donor agencies active in forestry and helped to mobilize additional funding for forestry. At the national level, the TFAP helped to focus attention on tropical deforestation, and stimulated interest among many governments in the preparation of actions plans designed to curb deforestation. The overall result has been an unprecedented commitment to promote the sustainable development of tropical forest resources backed up by a doubling of development assistance in forestry (FAO Forestry Department 1989c).

The planning sessions and meetings organized under the aegis of the TFAP served to promote dialogue between national governments and donor agencies and contributed to a shift in national policies and program priorities in a number of countries. In Peru, the government's desire to promote the expansion of industrial wood production in the Amazon basin was not supported by the aid agencies, although proposed projects in the area of community forestry and ecosystem conservation were endorsed at the donor's roundtable.

By encouraging the use of a fairly comprehensive framework, the TFAP process has helped increase the attention given to previously neglected subjects such as legislative reform and institutional restructuring. National forestry policies are being revised, forest concession management guidelines and fiscal policies are being reexamined, and other legislative and institutional changes are being proposed.

The TFAP planning process has helped to focus on the economic importance of the forestry sector. In several countries, the TFAP increased the attention given to the conservation of tropical forests in national development planning and in development assistance. In Zaire, Cameroon, Papua New Guinea, and other countries, the TFAP framework stimulated the preparation of projects aimed at "buffer zone" management around protected natural areas and other efforts to conserve the remaining forests.

Although it is still far from adequate, participation in several national TFAP exercised has been enhanced by the organization of roundtable meetings at various points in the planning process (Cort 1991). Roundtable meetings and workshops have been convened in Ecuador, Tanzania, Mali, Zaire, the Dominican Republic, and other countries to help inform NGOs and rural development organizations about the goals of the TFAP and to solicit their viewpoints and ideas on strategies and priorities actions to achieve these goals. In a few countries, small grants were provided to assist NGOs in the preparation of projects to be submitted as part of the national TFAP; such support for NGOs and organizations outside the government forestry department has been the exception rather than the rule, however.

Given the complex nature of deforestation and the challenge of sustainable use of forest lands, the most successful TFAP exercises seem to combine top-down and bottom-up approaches. Political commitment is needed from above to enact the policy reforms and to direct the flow of resources; simultaneously, local communities need to be involved in formulating realistic and appropriate strategies for improving their well-being. In this manner, government policies and programs support local-level initiatives, which in the aggregate make better use of the natural resource base without diminishing its long-term productivity.

Future Directions for the TFAP

A number of lessons have been learned to date in preparing and implementing national-level TFAPs. Several workshops and informal consultations that have been organized to assess the results of the TFAP suggest the following areas for improvement (Hazelwood 1987; Winterbottom 1990; WRI 1989, 1990b):

- Greater support should be given to broadly representative national steering committees and to country-level coordinators and researchers involved with the preparation of national TFAPs.
- A much broader range of disciplines and specialists should become involved in TFAP planning and implementation, including more social scientists, economists, and grass-roots development practitioners; such a multidisciplinary effort is needed to counteract the traditional bias of forestry professionals.
- There should be less dependence on government and development assistance agencies for information used in the TFAP planning process (especially regarding demography, land tenure, land use, proposed strategies, needed policy reforms, and action priorities); more support should be given to facilitate the contribution of other independent institutions, including universities, NGOs, and the private sector.
- More attention should be given to quality control in TFAP planning exercises and to the rigorous application of FAO guidelines for country-level TFAP exercises; more emphasis is needed on monitoring and evaluation of the TFAP.
- More attention should be focused on the special problems of the landless, rural poor, tribals, indigenous peoples, women, and other groups that are traditionally underrepresented in government and donor consultations.
- More attention should be given to underlying political and economic issues and to conflict resolution; this can be aided by improved information collection and analysis, easier access to TFAP documentation, broader dissemination of information, and increased support for dialogue among various interest groups.
- Greater emphasis should be given to noninvestment actions, especially in the early stages of the TFAP planning process, such as reallocation of development aid, information dissemination, and improved mechanisms to address multisectoral issues.
- The development assistance process should be streamlined and provide for the full participation of local communities and NGOs.
- Aid agencies and governments need to be more vigilant about addressing the real priorities for action and linking increased investment with needed policy reforms.
- The TFAP needs to be more closely coordinated with related initiatives in the area of conservation of biological diversity, debt reduction, and strategies to combat global warming and climate change.

Although the insufficient involvement of anthropology and other disciplines in the TFAP planning process in the past has compromised the results achieved, this deficiency is being increasingly recognized at both the international and national levels. A number of mechanisms are being proposed and tested to broaden participation in TFAP exercises and to ensure that the full range of important information and perspectives are factored into the preparation of national TFAPs.

Our experience with the TFAP has underscored the need to work closely with anthropologists and other social scientists in our efforts to improve the management of the world's forests. As we pursue the goal of "sustainable development"— development that meets the needs of people without destroying the natural resource base upon which we all depend—support is growing for a more people and policy-oriented approach to economic development, and for increased consideration of the socio-economic and ecological constraints to be addressed in development plans and programs. As Jim Nations, vice-president of Conservation International, has said: "If we don't look at the whole ecosystem, especially its human elements, everything else we do as conservationists is worthless."

References Cited

Bellagio Meeting on Tropical Forests
    1987 Statement of the Bellagio Strategy Meeting on Tropical Forests. Bellagio, Italy: Rockefeller Foundation, WRI, World Bank, UNDP, FAO.
Cabarle, Bruce J.
    1989 A Rumble from the Jungle: Plan de Accion Forestal Ecuador and Grassroots Participation: FCUNAE Workshop on Indigenous Peoples and Amazonian Forest Resources. Washington, DC: World Resources Institute.
Colchester, Marcus, and Larry Lohmann
    1990 The Tropical Forestry Action Plan: What Progress? London: World Rainforest Movement, The Ecologist, Friends of the Earth.
Cort, Cheryl
    1991 Voices from the Margin: NGO Participation in the Tropical Forestry Action Plan. Washington, DC: World Resources Institute.
Elliott, Chris
    1990 The Tropical Forestry Action Plan. World Wide Fund for Nature International, Gland, Switzerland.
Food and Agriculture Organization
    1982 Tropical Forest Resources. FAO Forestry Paper no. 30. Rome: FAO.
    1985 Tropical Forestry Action Plan. Committee for Forest Development in the Tropics. Rome: FAO.
Food and Agriculture Organization, Forestry Department
    1989a Guidelines for Implementation of the Tropical Forestry Action Plan at the Country Level. Rome: FAO.
    1989b Note on the Basic Principles of the TFAP. Rome:FAO.
    1989c Review of International Cooperation in Tropical Forestry, prepared for the Ninth Session of the Committee on Forest Development in the Tropics. Rome: FAO.
    1991 Tropical Forestry Action Plan Update No. 21. May. Rome: FAO.
Food and Agriculture Organization, World Resources Institute, World Bank, and United Nations Development Programme
    1987 Tropical Forestry Action Plan. Rome: FAO.
Halpin, Elizabeth
    1990 Indigenous Peoples and the Tropical Forestry Action Plan. Washington, DC: World Resources Institute.
Hazlewood, Peter T.
    1987 Expanding the Role of Non-Governmental Organizations in National Forestry Programs. Washington, DC: World Resources Institute and Environment Liaison Centre.
Lynch, Owen
    1990 Whither the People? Demographic, Tenurial and Agricultural Aspects of the Tropical Forestry Action Plan. World Resources Working Paper. Washington, DC: WRI.
Mahar, Dennis J.
    1989 Government Policies and Deforestation in Brazil's Amazon Region. Washington, DC: World Bank.
Myers, Norman
    1989 Deforestation Rates in Tropical Forests and Their Climatic Implications. Friends of the Earth Report. London.
Poore, Duncan, Peter Burgess, John Palmer, Simon Rietbergen, and Tim Synnott
    1989 No Timber without Trees. London: Earthscan.
Reid, Walter V., and Kenton R. Miller
    1989 Keeping Options Alive: The Scientific Basis for Conserving Biodiversity. Washington, DC: WRI.

Repetto, Robert
   1990   Deforestation in the Tropics. Scientific American 262(4):36–42.
Repetto, Robert, and Malcolm Gillis, eds.
   1988   Public Policies and the Misuse of Forest Resources. New York: Cambridge University Press.
Shiva, Vandana
   1987   Forestry Crisis and Forestry Myths: A Critical Review of Tropical Forests: A Call to Action. Penang, Malaysia: World Rainforest Movement.
Winterbottom, Robert
   1990   Taking Stock: The Tropical Forestry Action Plan after Five Years. Washington, DC: WRI.
World Resources Institute
   1985   Tropical Forests: A Call to Action. Report of an International Task Force convened by WRI, World Bank, and UNDP. Washington, DC: WRI.
   1989   NGO Consultation on the Implementation of the Tropical Forestry Action Plan, April 10–12. Washington, DC: WRI.
   1990a   World Resources 1990–91. New York: Oxford University Press.
   1990b   Report of a Workshop on Country-Level TFAP Exercises: Santo Domingo, Dominican Republic, October 24–26, 1989. Washington, DC: WRI.

Ecological Awareness and Risk Perception in Brazil

Alberto C. G. Costa
Conrad P. Kottak
Rosane M. Prado
John Stiles

We seek answers to several questions: How aware are people of ecological hazards of various sorts? Can and will they respond to them? Why do some people ignore real and evident threats to their health or well-being, while other people allow slight risks to inspire extreme fears? How is risk *perception* related to *actions* that could reduce threats to the environment and to health? Our research locale is Brazil, where three of the authors (Costa, Kottak, and Prado) are part of an international team (directed by Conrad Kottak of the University of Michigan) investigating environmental threats, their collective perception, and the growth of ecological awareness and action.[1]

Development and Environmentalism in Brazil

Brazilian economic development has proceeded in a society that has lacked generalized environmental awareness and regulation. With development pushed by both right and left, only in the 1980s did Brazilian politicians start emphasizing environmentalism. After long residence as exiles in Europe and North America, several leaders returned to Brazil with an environmentalist ideology. Today, Brazilian environmentalism is a growing political force, though still with largely urban support.

Generations of Brazilian politicians have advocated the emulation of European and North American patterns, including economic development. The earliest Brazilian industrialization, around São Paulo, can be traced to the 1920s. Industrialization continued with the overthrow of the established, largely agrarian, oligarchy by Getúlio Vargas (1937–1945). The emerging military and urban middle class favored state development of basic industries such as oil, steel, and hydroelectric power. The democratic regimes that followed Vargas, from the late 1940s through 1964, accepted his pro-development creed. Neither the military dictatorship installed in 1964 nor the subsequent *Nova República* (New Republic) veered much from it.

The Brazilian right has accepted "modernization theory" (Dahrendorf 1959) as a national ideology. Controlling the state for 25 years, it has justified "development" as a correct model of and for the relationship between Brazilians and their environment. The Brazilian left, though embracing "dependency theory" (Cardoso and Faletto 1979; Frank 1972) and proposing a different national trajectory, has also supported the development model. The left has viewed development as a condition of national autonomy and, when associated with a reduction in social inequality, as a solution to national problems.

Ecological awareness is most developed in Brazil's south-central and southern cities. It remains rudimentary elsewhere, despite widespread pollution and other environmental hazards. We offer a simple illustration from Arembepe (Bahia state), an Atlantic village that Kottak (1992) has been studying since 1962. Since the early 1970s, Arembepe has suffered air and water pollution from nearby factories. The municipal seat, Camaçari, has grown tenfold since 1964 and has become a petrochemical inferno. The region's streams, rivers, and coastal waters are chemically polluted, posing dangers to wildlife and people. Arembepeiros are truly at risk from pollution of the air, fresh water, and the ocean. Several times, reporters from the nearby metropolis of Salvador have covered Arembepe's chemical pollution. Although most villagers have seen those reports on TV, attention to local ecological problems has not increased concomitantly with the risks.

Walking on the beach near Arembepe in 1985, Kottak passed dead sea gulls every few yards—hundreds of birds in all. He watched the gulls glide feebly to the beach, where they soon died. Local people paid little attention. Stunned and suspecting a possible oil spill or mercury poisoning, Kottak asked for explanations, but people simply said, "The birds are sick."

Environmental Threats and Risk Perception

Ecological awareness among Brazilians began to grow in the mid-1980s for many reasons, including the return of public debate along with democracy. Since then, increased media coverage of ecological threats has raised risk perception. Especially significant was an accident involving radioactivity, which took place in the city of Goiânia, Goiás state, central Brazil, in September 1987. A diagnostic machine from an abandoned clinic was found by scavengers and sold to scrap metal dealers, who opened it. It contained a capsule of cesium 137 powder, which they also opened. The friends and family of the junk dealer handled the phosphorescent powder. Viewing the radioactive cesium as magical and possibly curative, they rubbed it on their bodies. All this became known when exposed people (more than 100, eventually) started showing signs of radiation sickness.

Other nationally publicized ecological threats followed. Most Brazilians have heard about the burning of the Amazon rainforest and the effects of road building, gold panning (using toxic mercury), and other intrusions of the world system on native groups and their lands. The media have also reported risks posed by oil spills, riverine mercury, industrial pollution, and poor waste-disposal, and the murder of the environmentally minded labor leader Chico Mendes. National ecological awareness grew before and during the Earth Summit, or UNCED (the United Nations Conference on the Environment and Development), held in Rio de Janeiro in June 1992. Concern with ecology has abated since then as the national focus has shifted toward politics and world soccer supremacy.

Our research project has investigated Brazilian ecological awareness and action through field studies at several sites facing different kinds and degrees of environmental threat. We are testing the proposition that risk perception does not arise inevitably from rational analysis and that a perception of risk does not guarantee action to remove the risk. We propose that risk perception emerges from encounters involving local ethnoecologies, imported ethnoecologies (often spread by the media), and changing circumstances (including population growth, migration, and industrial expansion).

Ethnoecologies

"The environment" becomes meaningful to people as they deal with it and construct cultural models (ethnomodels) of it. Every society has an *ethnoecology*—a traditional set of environmental perceptions and a cultural model of the environment in relation to people and society.[2] We may distinguish between *traditional ethnoecology,* an ethnomodel of *developmentalism,* and an ethnomodel of *environmentalism.* (We recognize that ethnoecologies include Western and scientific models along with native or local models, and that new international models, such as "sustainable development," have emerged. However, for simplicity, we restrict "ethnoecology" to the native models.)

Increasingly, traditional ethnoecologies are being challenged, transformed, and replaced through the modern world system as migration and industrialization relocate people and diffuse institutions. Introduced values and practices often conflict with those of natives. Consequently, in the contemporary context of population growth, migration, and commercial expansion, ethnoecological systems that have preserved local and regional environments for centuries are increasingly ineffective.

Also challenging traditional ethnoecologies are two, originally Euro-American, ethnoecologies—developmentalism and environmentalism. These models have spread as city, nation, and world increasingly invade local communities. The modern world has spawned immigrants, tourists, development agents, miners, ranchers, loggers, government and religious officials, politicians, the mass media—even rock stars promoting planetary ecology.

These external pressures enter myriad cultural settings, each of which has been shaped by particular national, regional, and local forces. (In our usage, "cultural" encompasses political economy.) Because different host communities have different histories and traditions, the impact of external forces is not universal or unidirectional. The spread of either developmentalism or environmentalism[3] is always influenced by national and local ethnoecologies and their powers of adaptation and resistance.

International groups try to impose their ideologies on nations and their people. (Many Brazilian nationalists perceive, in the first world environmentalist agenda, a hypocritical [at best] threat to their national sovereignty.) And national ethnoecologies increasingly challenge local ones. In one of the Brazilian field sites we are studying—Ibirama (Santa Catarina state)—federal regulation of the traditional timber industry began in October 1990. Elsewhere in Brazil, the population of Rondônia state has swollen through migration planned by the government and a highway financed by the World Bank. The planners' failure to predict and monitor the environmental impact led to soil exhaustion and devastation of Rondônia's Amazon forest. In many areas around the world, local peoples are being asked or forced to give up traditional hunting, gathering, herding, horticulture, and collecting either to make way for "development" or to allow parks and reserves to be established to satisfy the international environmentalist goal of "preserving biodiversity" (Kottak and Costa 1993).

Environmentalism entails a political and social opposition to the depletion of natural resources.[4] This concern has arisen in reaction to the expansion of a cultural model (developmentalism) shaped by the ideals of industrialism, progress, and (over)consumption. Environmental awareness is rising today as part of a change in which local groups adapt to new circumstances and to the models of developmentalism and environmentalism. Various threats spawned by development have been necessary conditions for the emergence of new perceptions of the environment.

We have chosen Brazil for an investigation of these matters because it illustrates a series of issues now being considered by scientists, governments, environmentalists, and the public. Our research has empirical and policy objectives, but it also asks theoretical questions: To what extent does risk perception arise from objective reasoning about real threats? To what extent is risk perception affected by political, economic, and other sociocultural factors? To what degree can the presence or absence of risk perception be explained by (1) cost-benefit analyses of risks, (2) psychological mechanisms of response to threats, and (3) cultural and social conditions?

Although an ecological hazard is necessary for environmental risk perception to emerge, the nature and gravity of that hazard do not explain the type of reaction that follows. Threats may be tolerated or avoided, emphasized or ignored. Risk perception emerges or languishes in *particular* cultural contexts. Recognizing cultural variation, it is important to consider that (1) people only react to threats they perceive; (2) risk perception is selective; (3) sets of values determine the perception of risks; (4) values are culturally and politically determined; and (5) the global spread of developmentalism and environmentalism is a political and economic process that entails cultural negotiation.

Risk Analysis Theory

Our approach challenges the prevailing sociological and psychological stances on risk perception as embodied in "traditional risk analysis" and "environmental psychology." The approach we label *traditional risk analysis*—mainly a product of sociology and social psychology—is aimed more at the public-policy goal of protecting people by assessing "real risks" (environmental hazards) than at understanding and predicting risk perception in a cultural context.[5] Traditional risk analysis studies groups directly exposed to potential hazards, such as nuclear facilities,[6] to produce a metrical or objective result, a "true measure" of risk.

The outcome of this approach often reveals a contrast between "expert risk assessment" and "public risk perception." Public perception, if it differs from the experts' conclusions, may be seen as a kind of illusion based on misleading information, misunderstanding, and irrationality. This can remove public (native) evaluations of risk from the formal analysis of risk and—a more serious problem from a policy standpoint—from the programs devised to respond to environmental threats.

Another aim of traditional risk analysis is to measure risk acceptability, focusing on the cognitive processes that shape responses to hazards. The main suggestion is that people react to such threats by calculating the costs and benefits of exposure. Sometimes cultural values are seen as important in these calculations; more often, self-interest is assumed to be the motivating factor.

A related belief—that environmental awareness and action will increase merely through risk perception—has dominated the Brazilian media coverage of environmental issues. The media follow traditional risk analysis in ignoring the role of culture in informing public perception of risk, mediating response, and creating awareness (see Wolfe 1988). The media often seem to consider local knowledge only as an obstacle to the emergence of ecological consciousness.

This attitude is even more characteristic of some works in *environmental psychology* (also known as prospect psychology). These works claim that culture is irrelevant in risk perception because people in hazardous environments "naturally" tend to minimize the danger and ignore warnings. This supposedly illustrates the role of "inferential biases" in helping people who face a threat to construct a predictable and safe environment. As with traditional risk analysis, the claim again is that risk assessment reflects basic cognitive processes, not culture.

The argument that culture is irrelevant to perception does not convince us. The often-cited conclusions of Tversky and Kahneman (see Kahneman et al. 1982) have been challenged within environmental psychology (see Berkeley and Humphreys 1982; Edwards 1983; Hogarth 1981; Sjöberg 1987). The narrow research universe in which Tversky and Kahneman worked makes the universality of their findings questionable. (Most of their interviewees were college students at the University of Oregon, whose "basic cognitive processes" are hardly independent of culture.) We maintain that cross-cultural comparison is essential to assess the universality of the "basic cognitive processes" supposedly identified by Tversky, Kahneman, and their associates.

Viewing reaction to environmental hazards in the context of varying cultural values can complement and broaden traditional risk analysis. Many risk analysts join us in emphasizing values and beliefs as reference points in studying public risk perception. Dubbed "postmodern risk analysis" (Rappaport 1988),[7] this approach, like the traditional one, examines groups exposed to environmental threats but sees values as crucial in risk perception, environmental consciousness, and individual and social action.

Our research agenda adds one more analytic level—*intercommunity comparison.* We view as incomplete any approach that studies only "at-risk groups"— people who live in hazardous environments. *Such an approach inevitably sees risk perception as a response to an immediate stimulus.*

We have chosen our field sites to represent a range of exposure to environmental hazards.[8] We assess the relative impact of cultural (including political and economic) factors, compared with objective threats, on risk perception and ecological awareness. We have learned that there are many reasons for variation in risk perception in Brazil. They include the existence of a multitude of local ethnoecologies within the vast nation; nature and degree of the environmental threat; level of education; socioeconomic status; and access to resources and information. People's values also influence their perception of environmental hazards and their actions.

The Importance of Cultural Values

The rise of ecological awareness is part of a larger process of cultural transformation. In this political process, hazards arising from a policy of national development may either favor or impede the emergence of (1) governmental policy concerned with environmental protection as a public responsibility and (2) enhanced awareness among Brazilians that protecting the environment is also a private responsibility.

Distrust of the public domain, as described by the Brazilian anthropologist Roberto DaMatta, is important here. DaMatta's book *A Casa e a Rua* (*The House and the Street*) (1987) delineates domains of social space recognized in Brazilian culture, running from home, through street, to the wider public world. DaMatta argues persuasively that these categories help us understand how Brazilians feel about issues of public responsibility. Brazilians' primary obligations lie within the house, broadly defined to encompass friends and the extended family. This is a personal domain whose social relations entail warmth, hospitality, trust, loyalty, and responsibility.

By contrast, the street and the nation-state are regarded as impersonal, disorderly, suspect, and inhumane. In DaMatta's view personal society (*casa*) is pitted against impersonal state (*rua, governo*). In the company of family and friends one is a *person,* but outside the casa one is an *individual,* a legal entity, a citizen—whose obligations may conflict with his or her primary social responsibility to family and friends. According to DaMatta, American culture creates no comparable opposition of person and individual (Americans are always individuals—even at home). Brazilians, by contrast, must shift between two codes of conduct—one for casa (person-society), the other for rua (individual-state). Primary respect and responsibility is reserved for the former. DaMatta discerns a much greater conflict between issues of private and public responsibility in Brazil than in the United States.

For example, Brazilians tend to be more suspicions about laws, enforcement, and government demands than Americans are, and Brazilians feel less responsible for the public domain than Americans do. The Brazilian house is supposed to be kept scrupulously clean, like the Brazilian person—which is bathed at least once a day. But one rarely encounters (as one often does in the Netherlands) someone washing the sidewalk in front of the home or "adopt-a-highway" programs like those in the United States. The streets, the beach, and the sea all belong to the public domain, for which the state, not people, is considered responsible.

These cultural contrasts—as described by DaMatta and extended here to the environmental arena—would seem to be differences of degree rather than kind. Most Americans also care more for their homes than for public space, and many Americans are suspicious of government. Still, we think that DaMatta has described an important area of contrast, significant to the cross-cultural understanding of environmental awareness and action. On the basis of his analysis, we would expect, for example, less voluntary compliance with environmental edicts and laws in Brazil than in the United States.

Recognizing the importance of culture—national and local—our central hypothesis is that environmental awareness is not a logically necessary response to perceived risks by individuals making objective and pragmatic judgments of reality. Sociocultural, economic, and political factors play a more important role than objective, intellectual, or scientific perceptions do. This is why environmental awareness and risk aversion may be most developed in social groups that are *least* threatened. (Compare a health-and-fitness-obsessed American yuppie with an impoverished fisherman from the highly polluted village of Arembepe.) We have found that Brazilian environmental awareness is most developed in the regions, communities, and social classes that are most directly influenced by international environmental concerns and ecological mobilization rather than in the places and groups most exposed to real threats.

The Social Context of Risk Perception

The first stage of our project (field research in the towns of Angra dos Reis and Ibirama) focused on a local setting influenced by Brazil's nuclear-energy program.[9] Angra dos Reis, Rio de Janeiro state, is a seaside resort and the site of Brazil's only nuclear reactor. In 1989 we originally chose Ibirama, a town in Santa Catarina state, as a control—to compare with the nuclear site. Ibirama faced no environmental threat evident to its residents. Now, however, Ibirama's traditional timber industry (in the Atlantic forest) is being affected by environmental legislation prompted by the national concern over deforestation. We are monitoring Ibirama's reaction to environmental measures imposed as of October 1990.

Statistical analysis of our quantitative data is incomplete, but we have reached some tentative conclusions. We found the people of "at-risk" Angra dos Reis to be much more aware of environmental dangers than the people of Ibirama were. Appendixes 1 and 2 list dependent variables (measures of risk perception—individual variables and indices, respectively) predicted by living in Angra dos Reis compared with the control community, Ibirama.

To assess the correlates and effects of exposure to an objective threat, versus sociocultural factors, on risk perception in Brazil, we used a standard set of eleven (potential) predictor variables. Using stepwise multiple regression, we examined their effects on several measures of risk perception (controlling in each case for the effects of the others). Our eleven predictors were *site, gender, age, skin color, social class, education, income, religiosity, print media exposure, length of home TV exposure*, and *current televiewing level.*[10]

In-depth interviewing, ethnographic fieldwork, and participant observation by Prado (1990) revealed that the people of Angra dos Reis from all social backgrounds had thorough knowledge of local, national, and international cases of environmental hazards. By contrast, Alberto Costa found that only upper-class Ibiramenses with higher levels of education and more exposure to the mass media had comparable knowledge.

These findings support the contention of traditional risk analysis that an environmental threat is crucial for risk perception and aversion. Our statistical analysis, particularly of the indices listed in Appendix 2, also show the people of Angra to have greater sensitivity to ecological issues and greater risk perception than the people of Ibirama do. *These findings fail to support the major postulate of environmental psychology—that people who face direct threats deal with anxiety by ignoring or minimizing risks.*

Still, the people of Angra dos Reis had not been convinced that nuclear energy, on which their town depends economically, is all bad. People in Angra were no more likely than those in Ibirama to reject the development model (e.g., by considering industry disadvantageous, saying people should not exploit nature, considering human use of nature disadvantageous, thinking natural is better, considering local life worse now than it used to be, or having a negative view of changes in the community). Nor were they more likely to reject nuclear energy (by disagreeing that nuclear energy is progressive and advantageous, or by seeing it as dangerous or as causing environmental pollution) (see Appendix 3).

Our research also confirms the prominent role of the mass media,[11] particularly television, in increasing risk perception. In our combined sample (Angra and Ibirama) HOMEEXPO (years of home television exposure) was the best predictor of the opinion that the community was dangerous,[12] that life is worse here now than it used to be,[13] and that changes in the community have been for the worse.[14] In Angra dos Reis HOMEEXPO was also the best predictor of a high score on the OCCULT index (see Appendix 2). This suggests that television may be playing a role in directing people toward Afro-Brazilian celebrations, spiritism, and New Age religion. This index may provide an indirect measure of risk perception.

Our research so far also suggests that risk perception does not necessarily correlate with participation in environmental organizations. In Angra dos Reis, although local people are well aware of environmental hazards, only the higher social classes join environmental organizations. People in Angra were no more likely than those in Ibirama to say they would like to participate in an ecological movement. They *were* more likely to say something should be done to prevent pollution and to know an ecology movement member.

In Ibirama, participation in environmental organizations has been restricted to a few upper-class members of the local branch of the Green party. This group's influence on local environmental awareness has been insignificant when compared with the role of television. The elitist character (and political insignificance) of Ibiramense environmental activism is revealed in local election results: The Green party candidate received less than 1 percent of the vote in the last (1988) race for mayor.

Field Research

By late 1993 field research had been completed at several sites: Angra, Ibirama, Arembepe, Goiânia, Cubatão, two Amazon sites, one shantytown (*favela*) in Rio de Janeiro, Caldas and Poços de Caldas in Minas Gerais state, and Aruaná in Goiás state. Two of our sites (Ibirama and one Amazon site) are in areas of deforestation. The other Amazon site faces mercury pollution, a strong threat to aquatic life. As mentioned, Arembepe (Kottak 1992) faces air and water pollution. Its people had little—but growing—ecological awareness when Kottak did field work there in 1991 and 1992.[15] Cubatão, São Paulo state, Brazil's most polluted city, has various chemical industries—including a huge oil refinery; it is internationally known as a symbol of air pollution and other hazards. It is also the site of a recent government program of environmental recovery. Goiânia had Brazil's only direct experience with a nuclear accident. Our range of research sites also includes controls—Brazilian communities that face no evident environmental threat. This range of sites should enable us to assess the impact of local cultural factors—versus objective hazards—on risk perception and ecological awareness. We believe our research design is useful and appropriate for an investigation of environmental awareness and risk perception because:

- Our field sites face different types of hazards.
- Our field sites face different degrees of ecological danger.
- Our field sites' economies depend on national and international markets to various degrees.
- Our field sites face different degrees of exposure to mass media and communications networks.
- The range of sites permits us to determine whether an objective environmental threat is a necessary and sufficient condition for risk perception and the emergence of ecological awareness.

Linkages Methodology

Our research design illustrates a "linkages" methodology. The linkages approach[16] accords with anthropology's traditional interest in cultural change. Its roots can be traced to earlier work, including Julian Steward's large-scale evolutionary and comparative projects,[17] the research of Max Gluckman and others who did "extended-case analysis," and world system approaches that emphasize the embeddedness of local cultures in larger systems.[18]

The linkages approach, as elaborated by Kottak and Colson (1994), agrees with world system theory that much of what goes on in the world today is beyond anthropology's established conceptual and methodological tools. Anthropologists must rethink their units of analysis and research methods. Traditional ethnography, based on village interviews and participant observation, assumed that informants knew what was going on in that delimited space. Today, however, no set of informants can supply all the information we seek. Local people may not be helpless victims of the world system, but they cannot fully understand all the relationships and processes affecting them.

Traditional ethnography also propagated the illusion of isolated, independent, pristine groups. By contrast, the linkages approach emphasizes the embeddedness of communities in multiple systems of different scale. Local people take their cues not just from neighbors and kin but also from a multitude of strangers—either directly or via the media. Anthropological research must expand to encompass linkages across levels of sociopolitical integration and across time and space.

Linkages research combines multilevel (international, national,[19] regional, local) analysis, systematic comparison, and longitudinal study (using modern information technology). Linkages methodology develops large-scale, explicitly comparative team projects (ideally involving international research collaboration). Linkages projects investigate aspects of change and development by blending ethnography and survey, synchrony and diachrony. Ideally, research is organized so that as new forces impinge on the study region, they can be examined in terms of their differential effects on known research populations. Dealing with social transformation, the linkages perspective considers both the exogenous pressures toward change and the internal dynamic of local cultures. As our current research demonstrates, we remain confident that a *fieldwork*-based anthropology can not only expand on the perspectives and findings of risk analysis but also contribute to a general theory of culture and cultural change.

Policy Implications

Throughout the world, local groups and their ethnoecologies face challenges from commercial expansion and development schemes. Other threats include population growth, immigration, and environmentalist-conservation-regulatory plans made regionally, nationally, and internationally. Environmental safeguards and conservation of scarce resources are important goals—from global, national, long-run, and even local perspectives. Still, ameliorative strategies must be implemented in the short run and in local communities. If traditional resources and products are to be destroyed, removed, or placed off limits (whether for development or conservation) they need to be replaced with culturally appropriate and satisfactory alternatives.

A new, possibly mediating, ethnoecological model, *sustainable development,* has emerged from recent encounters between local ethnoecologies and imported ethnoecologies, responding to changing circumstances. Sustainable development aims at culturally appropriate, ecologically sensitive, self-regenerating change. It thus mediates between the three models discussed in this article—traditional local ethnoecology, environmentalism, and developmentalism. We are investigating the extent to which the sustainable development model has reached Brazilians at different levels.

Experience designing the social-soundness component of the Sustainable and Viable Environmental Management (SAVEM) project, intended to preserve biodiversity in Madagascar, suggested that a gradual, sensitive, and site-specific strategy is most likely to succeed (Kottak and Costa 1993; Kottak and Rakotoarisoa 1990). Conservation policy can benefit from use of a flexible "learning process" model rather than a rigid "blueprint" strategy (Korten 1980; see also Kottak 1990a). The approach that Kottak recommended for Madagascar listens to the affected people throughout the whole process in order to minimize damage to them. Newly trained local socioeconomic experts—"para-anthropologists"—closely monitor the perceptions and reactions of the indigenous people during all the changes.

Because we, who include Brazilian and North American authors, believe that Brazil does need enhanced environmental awareness, we think our findings are relevant not only to academia but also to those (whether Brazilian or non-Brazilian) in a position to shape policy. We argue that the cultural approach we advocate will lead to better policy than will either traditional risk analysis or environmental (prospect) psychology.

Furthermore, research of the sort discussed here can help answer key questions for national (and global) environmental policy. There has been too little comparative local-level research on issues of risk perception, environmental awareness, and action. The answers are crucial for environmental preservation.

Notes

1. Pilot work for this research began in 1989 with funding from three agencies: the Michigan Memorial Phoenix Project (Project #714—The Social Context and Impact of Nuclear Energy in Brazil) and two Brazilian agencies—ANPOCS and Faperj.

2. Prior studies in cultural ecology have addressed the issue of the diversity of ethnoecological models and evaluated their adaptive or maladaptive functioning as a device of cybernetic regulation between culture and environment. See, for example, Bennett 1976, Ellen 1982, Frake 1962, Orlove 1980, and Rappaport 1984.

3. During the Industrial Revolution, a strong current of thought viewed industrialization as a beneficial process of organic development and progress. Many economists still assume that industrialization increases production and income. They seek to create in third world ("developing") countries a process like the one that first occurred spontaneously in 18th-century Great Britain. Development plans usually are guided by some kind of intervention philosophy, an ideological justification for outsiders to guide native peoples in specific directions. Bodley 1988 argues that the basic belief behind interventions—whether by missionaries, governments, or development planners—has been the same for more than 100 years. This belief is that industrialization, modernization, Westernization, and individualism are desirable evolutionary advances and that development schemes that promote them will bring long-term benefits to natives. We use *developmentalism* to describe ideology and practices that promote industrial development and *environmentalism* for ideology and practices aimed at conserving and preserving environmental resources and ecosystems.

4. Our operative definition of environmentalism is also sketched in the arguments of Bodley (1976:Chs. 1 and 2), Bramwell (1989:3–6), and Douglas and Wildavsky (1982:10–16).

5. As discussed Kasperson and Gray 1983 and Kasperson et al. 1988, "traditional" risk analysis seems to have as its primary aim the assessment of the "probability of events and the magnitude of specific consequences" rather than more comprehensive analyses that consider "social amplification."

6. For examples of "traditional risk analysis," see Hansson 1983, O'Riordan 1983, Paschen et al. 1983, Starr 1983, and Trunk and Trunk 1983.

7. For more culturally oriented approaches to risk analysis, see Covello and Johnson 1987, Douglas and Wildavsky 1982, Kasperson et al. 1988, Rappaport 1988, and Slovic et al. 1983.

8. This research has been supported by the National Science Foundation and by the Office of Forestry, Environment, and Natural Resources, Bureau of Science and Technology, of the U.S. Agency for International Development, under NSF Grant no. BNS-9112030.

9. Cooper 1985 provides support for our decision to begin our study of environmental risk perception with nuclear issues. Viewing radiation as the archetypal risk, he argues:

> The public tends to be particularly aware of disasters and potential disasters, and that is especially true of potential disasters from exposure to man-made radiation—from nuclear power, not from medical applications. The risks of radiation are particularly newsworthy and frequently come to the fore in general discussions of risk, for a variety of reasons, including its intangibility, its link with atomic bombs, and its links with cancer, birth defects, and heredity. [Cooper 1985:3–4]

10. Variable names within SPSS-PC are as follows: site (SITE), gender (INTSEX), age (INTAGE), skin color (COLOR), social class (LCLCLASS), education (INTEDUC), income (HHINCOME), religiosity (FREQCHUR), print media exposure (PRINT), years of home TV exposure (HOMEEXPO), and current televiewing level (ATVHOURS). For SITE, Angra dos Reis is coded 2 and Ibirama 1 so that correlations are positive when the people of Angra dos Reis have significantly higher risk perception measures than do people in the control community.

11. From 1983 to 1987, we also conducted research on the influence of the mass media in rural Brazil (see Kottak 1990b). Our national-level work was based on interviews and archival and statistical analysis. In our four local field sites, each differentially exposed to electronic media, ethnographic work provided the basis for representative sampling of households with differing degrees of access to television over time. Our findings, that length of home exposure predicted more dependent variables more strongly than did current viewing level, contrast with studies of TV's effects in Western culture, where length of home exposure is now effectively impossible to measure. We also found that although communication systems radiating from the major cities make people in the hinterlands aware of what they lack compared with city dwellers, especially urban elites, rural people do not passively accept the implicit teaching of these networks that they must change if they are to share in urban-based wealth. On the contrary, their comparisons of their lives with dominant metropolitan values are often favorable ones. These results, highly relevant to the study of the emergence of environmental awareness, confirm our belief in the power of the mass-media networks to have great influence on perceptions, and the notion that a negotiation of values takes places at local levels.

12. Initial Pearson's R and beta = .16656, R square = .02774.

13. Initial Pearson's R and beta = .20487, R square = .04197.

14. Initial Pearson's R and beta = .18173, R square = .03302.

15. Kottak's fieldwork in Arembepe in 1991 was facilitated by a grant to the University of Michigan from the National Aeronautics and Space Administration through the Consortium for International Earth Science Information Network.

16. This perspective was formalized at two Wenner-Gren–supported conferences organized by Douglas White and held in La Jolla, California, in 1986. Participants, who became founding members of Linkages: The World Development Research Council included Lilyan Brudner-White, Michael Burton, Elizabeth Colson, Scarlett Epstein, Nancie Gonzalez, David Gregory, Conrad Kottak, Thayer Scudder, and Douglas White.

Linkages' goals include assisting in organizing and coordinating basic scientific research on development on a worldwide basis. This includes formulating of theories, testing hypotheses, developing appropriate databanks for testing theoretical formulations, monitoring change, establishing trends, and identifying specific linkages or mechanisms involved in social change, including development interventions.

A crucial vehicle for development research, including study of both spontaneous and planned social change, is the systematic integration of data from longitudinal field sites. Such sites allow analysis and evaluation of long-term trends and effects, including cyclical changes relating to human populations and their ecologies, including the ecology of world systems and networks. Two such sites (Arembepe since 1962 and Ibirama since 1959) are included in the current project.

17. See *Area Research: Theory and Practice* (1950) and *The People of Puerto Rico* (1956). These books, cited much less often than his *Theory of Culture Change* (1955), have nevertheless influenced modern area studies and world system theory.

18. On the world system theory, see Comaroff 1982, Mintz 1985, Moore 1986, Nash 1981, Nugent 1988, Roseberry 1988, Schneider 1977, Wallerstein 1974, and Wolf 1982.

19. Our data on Brazilian environmental awareness come from the national level as well as from local field sites. Our national-level work focuses on federal archives, policy makers, the mass media, and environmental organizations. Some of the questions we are investigating include: How do the media represent technological and industrial development? How do they relate development to environmental pollution and ecological consciousness? How do environmental organizations shape public opinion, media coverage, official decisions, and policy in relation to ecological issues?

References Cited

Bennett, John W.
    1976 Anticipation, Adaptation, and the Concept of Culture in Anthropology. Science 192:847–853.
Berkeley, Dina, and Patrick Humphreys
    1982 Structuring Decision Problems and the Bias Heuristic. Acta Psychologica 50(3):201–252.
Bodley, John H.
    1976 Anthropology and Contemporary Human Problems. Menlo Park, CA: Cummings.
    1988 Tribal Peoples and Development Issues: A Global Overview. Mountain View, CA: Mayfield.
Bramwell, Anna
    1989 Ecology in the 20th Century: A History. New Haven, CT: Yale University Press.
Cardoso, Fernando H., and Enzo Faletto
    1979 Dependency and Development in Latin America. Berkeley: University of California Press.
Comaroff, John
    1982 Dialectical Systems, History and Anthropology: Units of Study and Questions of Theory. The Journal of Southern African Studies 8:143–172.
Cooper, M. G., ed.
    1985 Risk: Man-Made Hazards to Man. Oxford: Clarendon Press.
Covello, Vincent T., and Branden Johnson, eds.
    1987 The Social and Cultural Construction of Risk: Essays on Risk Selection and Perception. Dordrecht: D. Reidel.
Dahrendorf, Ralf
    1959 Class and Class Conflict in Industrial Society. Stanford, CA: Stanford University Press.
DaMatta, Roberto
    1987 A casa e a rua. Rio de Janeiro: Guanabara.
Douglas, Mary, and Aaron Wildavsky
    1982 Risk and Culture: An Essay on the Selection of Technical and Environmental Dangers. Berkeley: University of California Press.
Edwards, Ward
    1983 Human Cognitive Capacities, Representativeness and Ground Rules for Research. *In* Analyzing and Aiding Decision Processes. Patrick Humphreys, ed. Pp. 507–513. Amsterdam: North-Holland.
Ellen, Roy
    1982 Environment, Subsistence, and System: The Ecology of Small-Scale Social Formations. Cambridge: Cambridge University Press.
Frake, Charles O.
    1962 Cultural Ecology and Ethnography. American Anthropologist 64(1):53–59.
Frank, Andre G.
    1972 The Development of Underdevelopment. *In* Dependence and Underdevelopment in Latin America's Political Economy. James Cockcroft, Andre G. Frank, and Dale Johnson, eds. Pp. 3–17. New York: Anchor.

Hansson, Bengt
  1983 The Assessment of Nuclear Risk: Some Experiences from the Swedish Energy Commission. *In* The Analysis of Actual versus Perceived Risks. Vincent T. Covello, W. Gary Flamm, Joseph V. Rodricks, and Robert G. Tardiff, eds. Pp. 69–80. New York: Plenum Press.
Hogarth, Robin
  1981 Beyond Discrete Biases: Functional and Dysfunctional Aspects of Judgmental Heuristics. Psychological Bulletin 90:197–217.
Kahneman, Daniel, Paul Slovic, and Amos Tversky, eds.
  1982 Judgement under Uncertainty: Heuristics and Biases. New York: Cambridge University Press.
Kasperson, Roger E., and Arnold L. Gray
  1983 Risk Assessment Following Crisis in the United States: The Kemeny Commission. *In* The Analysis of Actual versus Perceived Risks. Vincent T. Covello, W. Gary Flamm, Joseph V. Rodricks, and Robert G. Tardiff, eds. Pp. 129–156. New York: Plenum Press.
Kasperson, Roger E., Ortwin Renn, Paul Slovic, Haline S. Brown, Jacque Emel, Robert Goble, Jeanne Kasperon, and Samuel Ratick
  1988 The Social Amplification of Risk: A Conceptual Framework. Risk Analysis 8(2):177–187.
Korten, David C.
  1980 Community Organization and Rural Development: A Learning Process Approach. Public Administration Review (September–October):480–512.
Kottak, Conrad Phillip
  1990a Culture and Economic Development. American Anthropologist 92(3):723–731.
  1990b Prime-Time Society: An Anthropological Analysis of Television and Culture. Belmont, CA: Wadsworth.
  1992 Assault on Paradise: Social Change in a Brazilian Village. 2nd ed. New York: McGraw-Hill.
Kottak, Conrad Phillip, and Elizabeth Colson
  1994 Multilevel Linkages: Longitudinal and Comparative Studies. *In* Assessing Anthropology. Robert Borofsky, ed. Pp. 396–412. New York: McGraw-Hill.
Kottak, Conrad Phillip, and Alberto Costa
  1993 Ecological Awareness, Environmentalist Action, and International Conservation Strategy. Human Organization 52(4):335–343.
Kottak, Conrad Phillip, and Jean-Aimé Rakotoarisoa
  1990b Social-Soundness Analysis for SAVEM—Sustainable and Viable Environmental Management, USAID-Madagascar. Unpublished report.
Mintz, Sidney
  1985 Sweetness and Power: The Place of Sugar in Modern History. New York: Viking.
Moore, Sally F.
  1986 Social Facts and Fabrications. Cambridge: Cambridge University Press.
Nash, June
  1981 Ethnographic Aspects of the World Capitalist System. Annual Review of Anthropology 10:393–423.
Nugent, Stephen
  1988 The Peripheral Situation. Annual Review of Anthropology 17:79–98.
O'Riordan, Timothy
  1983 Coping with the Risks of Nuclear Power Plants in the United Kingdom. *In* The Analysis of Actual versus Perceived Risks. Vincent T. Covello, W. Gary Flamm, Joseph V. Rodricks, and Robert G. Tardiff, eds. Pp. 101–128. New York: Plenum Press.
Orlove, Benjamin S.
  1980 Ecological Anthropology. Annual Review of Anthropology 9:235–273.
Paschen, H., G. Bechmann, and G. Frederichs
  1983 Nuclear Power Plant: West German Management of Risk: A Problem Analysis. *In* The Analysis of Actual versus Perceived Risks. Vincent T. Covello, W. Gary Flamm, Joseph V. Rodricks, and Robert G. Tardiff, eds. Pp. 81–99. New York: Plenum Press.
Prado, Rosane
  1990 Beauty Betrayed: Risk Perception at a Nuclear Reactor Site in Brazil. Paper given at the 89th annual meeting of the American Anthropological Association, New Orleans, LA.

Rappaport, Roy A.
   1984   Pigs for the Ancestors. Rev. ed. New Haven: Yale University Press.
   1988   Toward Postmodern Risk Analysis. Risk Analysis 8(2):189–191.
Roseberry, William
   1988   Political Economy. Annual Review of Anthropology 17:161–185.
Schneider, Jane
   1977   Was There a Pre-Capitalist World System? Peasant Societies 6:20–28.
Sjöberg, Lennart, ed.
   1987   Risk and Society: Studies of Risk Generation and Reactions to Risk. London: Allen
          & Unwin.
Slovic, Paul, Baruch Fischoff, and Sarah Lichtenstein
   1983   The Public vs. the Experts: Perceived vs. Actual Disagreements about Risks. In The
          Analysis of Actual versus Perceived Risks. Vincent T. Covello, W. Gary Flamm, Joseph
          V. Rodricks, and Robert G. Tardiff, eds. Pp. 235–249. New York: Plenum Press.
Starr, Chauncey
   1983   Coping with Nuclear Power Risks: A National Strategy. In The Analysis of Actual
          versus Perceived Risks. Vincent T. Covello, W. Gary Flamm, Joseph V. Rodricks, and
          Robert G. Tardiff, eds. Pp. 251–257. New York: Plenum Press.
Steward, Julian
   1950   Area Research: Theory and Practice. New York: Social Sciences Research Council.
   1955   Theory of Culture Change. Urbana: University of Illinois Press.
   1956   The People of Puerto Rico. Urbana: University of Illinois Press.
Trunk, Anne, and Edward Trunk
   1983   Impact of the Three Mile Island Accident as Perceived by Those Living in the
          Surrounding Community. In The Analysis of Actual versus Perceived Risks. Vincent T.
          Covello, W. Gary Flamm, Joseph V. Rodricks, and Robert G. Tardiff, eds. Pp. 225–233.
          New York: Plenum Press.
Wallerstein, Immanuel
   1974   The Modern World-System: Capitalist Agriculture and the Origin of the European
          World Economy in the Sixteenth Century. New York: Academic Press.
Wolf, Eric
   1982   Europe and the People without History. Berkeley: University of California Press.
Wolfe, Amy K.
   1988   Environmental Risk and Anthropology. Practicing Anthropology 10:3–4.

Appendix 1—Variables Predicted by Being "At Risk"[1]

NATSC19   Agrees that people always destroy nature by *exploraçaõ*[2]
FILTER   Has a water filter
DANGER1   Considers the world dangerous
DANGER2   Considers the community dangerous
DANGER5   Considers the community more dangerous than it used to be
NCLEAR1   Has heard of nuclear energy
ENVIRO5   Is worried about environmental pollution
LCLENV1   Recalls local or personal risk from environmental pollution
LCLENV3   Thinks community is polluted like a big city
LCLENV5   Thinks community is more polluted now than in the past
ENVEGO1   Says he or she has done something to pollute
ENVPREV1   Thinks something should be done to prevent pollution
ECOMOV3   Knows an ecology movement member
WRLEND1   Thinks the world might end
GOIANA4   Thinks Goiânia posed an immediate danger
PANTAN1   Has heard of Pantanal pollution
CHERNO1   Has heard of Chernobyl

---

[1]That is, people in the nuclear reactor site, Angra dos Reis, had significantly higher scores on these risk perception measures than did people in the control community, Ibirama.
[2]Brazilian Portuguese uses the same word for explore and exploit.

Appendix 2—Indices Predicted by Being "At Risk"[1]

EXTRALOC	enviro1+cherno1+goiana1+amazon1+pantan1
SENS	enviro4+enviro5+lclenv1+envego1+envoth1
EXTAWARE	EXTRALOC+SENS
INFORM	enviro1+nclear1+cherno1+goiana1+amazon1+pantan1+angra1
AWARE	INFORM+SENS
DANGER	danger1+danger2+danger5
RISK	DANGER+wrlend1+goiana4
OCCULT	spcenter+umbanda+curer+diviner

Definition of Variables

ENVIRO1	Have you heard of environmental pollution?
CHERNO1	Have you heard of Chernobyl?
GOIANA1	Heard of Goiânia accident?
AMAZON1	Have you heard about Amazonian deforestation?
PANTAN1	Have you heard of Pantanal pollution?
ENVIRO4	Interested by info about environmental pollution
ENVIRO5	Are you worried about environmental pollution?
LCLENV1	Recalls local or personal from environmental pollution
ENVEGO1	Do you or have you done anything to pollute?
ENVOTH1	Knows a polluter
DANGER1	Is the world dangerous?
DANGER2	Is your community dangerous?
DANGER5	Is your community more dangerous now than it used to be?
WRLEND1	Do you think the world might end?
GOIANA4	Did Goiânia pose any immediate danger?
SPCENTER	Goes to spiritist center
UMBANDA	Goes to terreiro de umbanda
CURER	Goes to benzedeira
DIVINER	Goes to sortista

---

[1]That is, people in the nuclear reactor site, Angra dos Reis, had significantly higher scores on these risk perception measures—indices summing several questions—than did people in the control community, Ibirama.

Appendix 3—Variables Not Predicted by Being "At Risk"[1]

LCINDS5	Considers industry disadvantageous for the community
NATURE9	Considers nature dirty
NATURE10	Considers nature disorderly
NATCHAOS	Considers nature dirty and disorderly (NATURE9+NATURE10)
NATSCI3	Disagrees that people have the right to *explorar* nature
NATSC17	Disagrees that progress depends on the *exploraçao* of nature
NATSC18	Considers human *explorçao of nature disadvantageous*
NATSCI10	Disagrees that science can *explora*nature nondestructively
NATURE18	Thinks natural is better than nonnatural
LCCHAN1	Considers local life worse now than it used to be
LCCHAN4	Negative options of changes in the community
NCLEAR4	Disagrees that nuclear energy is progress
NCLEAR5	Disagrees that nuclear energy is advantageous
NCLEAR6	Agrees that nuclear energy is dangerous
ENVIRO4	Is interested by information about environmental pollution
NCLENV	Agrees that nuclear energy causes environmental pollution
ENVOTH1	Knows someone who pollutes
ENVPREV2	Agrees that something should be done to prevent pollution now
ECOMOV4	Would like to participate in ecological movements
GOIANA1	Has heard of the Goiânia accident
AMAZON1	Has heard about Amazonian deforestation
SOLVE1	Disagrees that problems at Angra have been solved
EXPLOIT	Index: NATSC13+NATSC14+NATSC16+NATSC17+NATSC18
PRINT	Index: of print media exposure

Definition of Variables

NATSC13	Do humans have right to *explorar* nature?
NATSC14	Do humans need to *explorar* nature?
NATSC16	Do humans render nature productive?
NATSC17	Does progress depend on the *exploraçao* of nature?
NATSC18	Is human *exploraçao* of nature advantageous?
LITERACY	Knows how to read and write
NEWSPP1	Reads newspapers
MAGAZ1	Reads magazines
BOOK1	Reads books

---

[1]That is, there was no statistically significant difference in responses of people in nuclear reactor site, Angra dos Reis, and people in the control community, Ibirama.

The Cultural Environment of Development: Commentary

Terence Turner

The historical difficulties experienced by anthropologists in combining their roles as anthropologists with participation in practical projects for social and economic change are symptomatic of fundamental problems in the theory and practice of anthropology as a discipline and, to an equal or greater extent, in the theory and practice of development. As Painter succinctly observes in his article, anthropologists have responded to the challenge of "applying" their professional knowledge and skills either by "retreating into the academy and pretending to deal with [theoretical] issues that transcend" such practical concerns or

> plung[ing] headlong into the service of a wide range of institutions, eschewing the relevance of a large part of anthropological theory to our [i.e., applied anthropologists'] work, and promoting an emasculated series of data-gathering techniques as a supposedly value-free applied anthropology.

Painter's point, however, might be put more strongly by taking account of the fact that the models, ideals, and methods of development presupposed by the great majority of applied anthropology projects have been such as to preclude consideration of the non-Western cultures, social forms, and values with which anthropological theory has been chiefly concerned as anything more than obstacles to be overcome. The goals and analytical perspectives of most applied anthropology have been defined not so much by anthropology as by the mixture of neoclassical economics, political science, sociology, and various Western technologies (civil engineering, medicine, agriculture, etc.) common to what are now often referred to as the public policy disciplines, with anthropology as such coming in only at the level of intercultural translation, explaining the natives to the developers and the developers and their programs to the natives.

Painter is nonetheless correct that anthropology as a theoretical discipline has failed on the whole to provide either a practical or a critical framework for orienting the participation of anthropologists in development activities involving the peoples with whom they have worked. Applied anthropology has historically been regarded as a stepchild of the discipline, its activities and concerns falling outside the major theoretical foci of academic anthropology. The problem has not been so much that anthropological theory transcends issues of practical social application as that the applications with which applied or development anthropology are concerned have typically been oriented away from the sphere of indigenous cultural, social, and adaptive forms toward integration in Western or national third world economies and institutional orders, and conceived from the perspective of those orders rather than "from the native point of view."

For its own part, academic anthropology has largely failed to integrate its theoretical notions of culture and social structure with critical sociological and political-economic perspectives. An influential subset of the discipline, carrying on Boasian and Weberian orientations and continuing, with only slight modification, in various postmodern forms, has persisted in defining culture in terms of "systems of symbols and meanings" (or, in currently trendy terms, "discourses," "texts," or "writing") that lack any analytically specifiable relation to the social and material conditions they mediate. Such approaches provide little basis for practical social or economic "applications" of any kind. Thus, although it is doubtless true, as Painter and Gow charge, that many anthropologists have been content to "retreat into the academy," it is important to recognize that, given the theories of culture with which considerable numbers of them have been operating, they had left themselves little choice.

Both sides of the problem of the disarticulation of anthropological theory from applied or development anthropology may be seen to derive from the same fundamental assumption, namely, the incommensurability of non-Western and Western cultures and, by extension, the irrelevance and unviability of non-Western cultures, especially those of relatively "primitive" peoples lacking economies based on commodity production and class exploitation, in the modern capitalist world system. This assumption, in turn, can be understood as an integral part of the hegemonic ideology of Western world domination and the triumph of industrial capitalism that has served, with minor variations in the state socialist countries, as the context of the historical development of anthropology.

The theories of development and modernization that applied anthropologists sought to apply, drawn as they were largely from economics, sociology, and political science, precluded any significant role for anthropological notions of culture or the value of indigenous forms of social organization. In the perspective of neoclassical development economics or neo-Weberian modernization theory, the cultures and social institutions of the peripheral peoples and local communities who were the objects of development could only appear as obstacles to be reformed as part of the development project. Lacking as they invariably do Western institutional forms of private property, profit and accumulation, and economically "rational" patterns of saving and investment, they could not be seriously considered as potentially positive factors in the struggle for material and social empowerment oriented to locally meaningful values. The role of the anthropologist, given these assumptions, was too often confined to that of translator of imposed Western ideals into the local cultural idiom they sought to supplant, and facilitator of the dismantling of local social structures and the values they embodied.

Perhaps the most telling manifestation of the pervasiveness of the hegemonic assumptions of inevitable Western capitalist domination and the lack of a historical future or developmental potential of non-Western cultures was the tacit convergence of academic and development anthropologists on these fundamental points with a third group of anthropological thinkers and activists who usually defined themselves overtly in opposition to them both, namely, the exponents of "Indigenismo" in its traditional Latin American forms. Indigenist thinkers and activists, like old-style development anthropologists, tended to see the cultural values and social formations of non-Western peoples as historical dead ends, insofar as they differed or clashed with those of the enveloping national society, total assimilation into which tended to be taken for granted as the ultimate goal. That these same native cultural values and social forms might constitute, not obstacles to development or effective political resistance, but more viable forms of social production, ecological accommodation, and political defense of local autonomy thus seems until recently to have been as rarely imagined by many indigenist advocates as by applied/development anthropologists. The resulting paternalistic forms of Indigenismo, with Western indigenists cast in the role of protectors and (relatively gentle) assimilators of culturally incompetent natives, may be seen in a critical historical perspective as sharing many of the assumptions of the applied and development anthropological approaches of the 1950s, 1960s, and 1970s (Turner 1993).

In the late 1960s and early 1970s, a combination of factors had begun to undermine and discredit these aspects of the conventional paradigm of applied anthropology. Probably most important were the reactions against U.S. foreign policy, including its aid programs, aroused by the Vietnam War, and the shift of U.S. aid policy away from its admittedly tenuous committment to promoting local democracy and community self-betterment to counterinsurgency and the repression of popular resistance movements associated with that war. The politicization and radicalization of theoretical orientations within anthropology brought about by the war also contributed to a more critical view of the neoclassical economic ideals of development and the political-economic effects and interests served by Western-financed development projects. The rise of the environmentalist movement contributed another critical perspective on the effects of development projects on local regions and the people who inhabit them. A series of scandals involving abuses of the human rights, lands, and lives of native peoples (such as the Amazonian Indians) and peasant populations, and the resulting growth of native peoples' and peasant resistance movements (such as the independence struggles in central and southern Africa, the smoldering revolt in the Phillipines, and the mass mobilization of peasants and tribal peoples against dams and other projects in India) was another powerful factor in shifting the focus of concern of many anthropologists from liberal models of development assistance to the defense of rights and the support of resistance.

All of these factors led to a phenomenal growth in the number and influence of nongovernmental organizations (NGOs) dedicated to environmentalism, indigenous advocacy and human rights, and neophilanthropic relief and support groups (such as Oxfam) aimed at populations often victimized as much by political and social oppression as by poverty, disease, or natural disasters. These groups provided an alternative venue for anthropologists committed to working in applied or practical capacities on problems related to development. For anthropologists influenced by or working in such groups, or sharing similar views and approaches, the relatively economistic and administratively top-down notions of the development planners, engineers, and agronomists of the 1950s and 1960s tended to give way to more politically radicalized approaches, stressing more oppositional relations to first and third world governments and international development banks, support of popular resistance and the rights of oppressed indigenous and ethnic minorities, and the protection of environmental values: in sum, a more critical theoretical perspective on the nature of the problems and forms of development to be addressed.

During the 1980s these radicalized views and the policies and orientations of the NGOs that sought to put them into effect exercized an increasing influence over the views and activities of many anthropologists working in applied anthropological or development projects, intergovernmental aid programs, and locally initiated, community-level projects. This hybridization between conventional applied or development-oriented anthropology, on the one hand, and approaches developed or influenced by advocacy NGOs and indigenous support politics, on the other, is exemplified by most of the articles in this volume. Case reports and programmatic formulations exemplifying this tendency represent at once the most distinctive common theme and the most significant contributions of the volume.

The impact of this historic convergence of influences from advocacy and political support groups on applied and development anthropology as a field may be summed up as a change on the part of many of the anthropologists involved in the idea of what applied anthropology should be applied *for* and of what should be developed by development. As illustrated by Costa et al.'s analysis of environmental awareness and activism in Brazil, the reinforcement or protection of local political autonomy, subsistence resource base, cultural self-determination, and environmental quality are increasingly prominent as goals of applied anthropological practice or as criteria of evaluation by applied anthropologists of the impact of existing or contemplated projects. Anthropologists working in the general field of development are now much more likely to think in terms of ecological conservation, local political control, and retention of economic benefits, and even the promotion or reinforcement of local communities' resistance to development schemes of central governments and international financial institutions damaging to their interests and environment.

Just as applied and development anthropologists have learned from the work of environmental advocacy and indigenous support groups, so the latter have had to learn much about the techniques, policy expertise, and developmental theory of the former. The fertilization of the two fields has thus run in both directions. To be sure, advocacy NGOs have been more concerned to criticize and oppose the sorts of development projects and policies with which applied and development anthropologists have tended to be identified, but they have also become involved in various kinds of applied projects of their own. Some of the most effective applied anthropology of the past decade has been done by anthropologists working either through NGOs or directly with local peoples in organizing resistance to development schemes that promised to have disastrous environmental, social, and cultural effects within the local area of their implementation.

By way of illustration, I might mention a few examples of this sort with which I have had the good fortune to be involved. One that comes readily to mind is the successful campaign of the Kayapo Indians of Central Brazil, supported by several anthropologists and both environmentalist and indigenous advocacy NGOs, against the hydroelectric dam scheme planned by the Brazilian government for the Xingu River valley, for which that government had applied for a World Bank loan. A second is the Video in the Villages project of the Brazilian advocacy organization, the Centro de Trabalho Indigenista, which carries out video documentation of Brazilian indigenous groups and their activities at the request and under the supervision of the indigenous communities themselves. The more recently founded Kayapo Video Project, which provides video cameras and training in camera use and editing to the Kayapo, has been supported in impotant ways by Video in the Villages and is close to it in orientation. These two loosely affiliated projects have had a significant impact on cultural self-conscientization, political activism, and ethnic identity-formation by a number of indigenous Amazonian communities.

A third example of a different kind is afforded by the the campaign to bring about Brazilian government recognition and police protection of the Yanomami territory in Northern Brazil, which finally led to the demarcation of an integral Yanomami Reserve containing a reasonable approximation of the full tribal territory in November 1991. Various Brazilian NGOs such as the Commission for the Creation of a Yanomami Park (CCPY), the Ecumenical Center for Documentation and Information (CEDI), and Action for Citizenship (Açao pela Cidadania), international indigenous advocacy NGOs such as Survival International and Cultural Survival, and professional anthropological organiations such as the Associação Brasileira de Antropologia and the American Anthropological Association played important parts in this long and ultimately successful struggle. This case affords a dramatic example of the way that involvement in advocacy and support roles in defense of indigenous rights has become an increasingly important form of applied anthropological activity not only for individual anthropologists but even professional anthropological organizations such as the American Anthropological Association.

A characteristic of the approach of the older applied anthropology of the 1950s and 1960s was its tendency to overlook conflicts of interest and structural contradictions among those affected by development projects and often exacerbated by them. This can be recognized as a consequence of uncritically accepting functionalist and positivist notions of value-free data collection and methodology associated with categories such as "development," "underdevelopment," "acculturation," and the like.

One of the pervasive sites of conflict is the contradictory relation between local communities, with their aspirations for relative autonomy and economic betterment, national governments, with their ambitions for centralized control and capital accumulation, and international capital, particularly as represented by the aid programs of first world governments and international development banks, which has consistently pursued policies that have promoted the ever-deepening debt bondage of much of the third world, with disastrous effects on both governmental and local autonomy and economic progress. Another point of conflict has been the destructive effects of much development on ecological systems, and another has been the exacerbation of gender and class inequalities in local communities. These foci of conflict have also been the main foci of conflict and contradiction in the approaches of nongovernmental advocacy groups on the one hand and applied/development anthropologists on the other, with advocacy NGOs typically taking the side of local communities and minorities against national governments and international forces, and attempting to defend the natural environment from damage by developers.

The inevitable consequence of some applied and development anthropologists' taking on critical views and approaches convergent with those of advocacy NGOs is the importation of these contradictions and conflicts into the fields of applied and development anthropology in the form of sharpened incompatibilities between community-oriented and nationally or internationally oriented approaches. An example of this tendency in this volume is the contrast between the approaches of Wright and Winterbottom to problems of ecological conservation, the former focused at the local community level, the latter focused at the national and international levels. As I read Winterbottom's article, the main source of difficulties of the international Tropical Forestry Action Plan is that it is formulated at a level that fails to address the sort of local-level factors addressed by the hearteningly successful projects described by Wright. Not that a focus on the local level is sufficient in and of itself: Painter's analysis is an excellent example of the mutual dependence (and mutual contradictions) of local and national/international conditions, developments, and policies.

Given the contradictory aims and needs of local communities, nation-states, and international capital, and, at times, of any or all of the three with the natural environment, it will of course continue to be impossible to develop integrated approaches in development anthropology that reconcile all levels and types of factors. By the same token, for all their mutually beneficial convergences, advocacy anthropology and applied/development anthropology will remain essentially separate and distinct approaches to similar phenomena. In sum, applied anthropology, like the real world, will perforce remain a mass of contradictions, but as the contributions to this volume demonstrate, at least the discipline is learning to confront and encompass more of the real contradictions that in its earlier, more formative stages were ignored or left implicit.

This point is neatly exemplified by the juxtaposition of the three articles to which I have just referred. Wright's, Painter's, and Winterbottom's articles form a set, all three exemplifying the convergence of environmentalist and critical political perspectives with an anthropologist's concern for formulating programs in terms of local needs and values so as to maximize local control and leadership. All three clearly recognize the emptiness of a purportedly value-free emphasis on local participation that fails to recognize that the locals are divided by class, gender, and mode of subsistence (e.g., subsistence farmers, ranchers, loggers, miners, and oil men), such that any program of development must promote the interests of some participants more than others.

The most important point about these three articles, however, is that they go beyond an abstract theoretical recognition of this point to demonstrate how it can be taken into account in principled ways in concrete attempts to promote the interests of the local populations who have so regularly been the losers in centralized, capital-intensive development projects. Wright's Zambian example of how a game conservation program based on the cooperative participation of local subsistence farmers, planned in collaboration with local leadership, can avoid the manifold social problems and rampant poaching associated with the game-park approach of other African countries, is particularly apposite in this regard. His emphasis on the point that local participation must mean not only a selective identification of which locals are to participate but also careful attention to the effects of the project and its projected benefits on local social and political patterns (and, specifically, how these might entail conflicts with or reinforcement of the interests of local leadership, gender relations, etc.) is well justified by his example. Implicit in his account of the planning of the Zambian program is the point that the anthropological consultant who engages at this level cannot operate in a value-free way. He or she must willy-nilly become an advocate of some interests against others. In Wright's case, the decision was taken to support the interests of existing local leadership and the local social order of subsistence farmers in order to provide sufficient local incentives to the local population to induce them to support the program. As Wright himself recognizes, however, this meant favoring men's interests over women's, among other ways by "favoring the male domination of community decision making" (this despite the fact that "women are the primary users of the natural resource base" affected by the project).

Here we find an unusually clear-eyed confrontation of the limiting contradiction of an applied anthropology that commits itself in advance to the promotion of development within existing social and political conditions. To achieve success in these terms, and even to gain permission from local authorities to make the attempt, development anthropologists must work within existing social and political structures; their programs must therefore have a built-in conservative bias. Wright's case is a good example of how it is possible to work within these limitations to achieve some progressive social and political goals, as well as more effective environmental conservation. The dilemma, however, remains, as Wright succinctly points out.

Dilemma becomes contradiction when the progressive political values that guide the emphasis on local popular participation and benefits in programs such as Wright's Zambian case lead on toward promoting the interrests of relatively disadvantaged popular elements (e.g., women) that would disrupt the interests of the dominant elements on whose collaboration the success, and indeed the existence of the program depends. Here, one reaches the limit that continues to divide development anthropology, as an essentially liberal project, from advocacy anthropology, to employ this term for more radical anthropological approaches that deliberately engage themselves in political struggle on behalf of oppressed groups (whether defined in cultural, gender, ethnic, or political-economic terms) in ways that entail direct challenges to existing structures of dominance, in an effort to transform them.

It is important for advocacy anthropologists to recognize that such struggles cannot be fought only on the boundaries between ethnic or cultural minorities and national societies, states or world systems; they must also be engaged *within* indigenous and other traditional minorities or oppressed groups themselves, to the extent that they oppress groups or categories of their own members. The "existing structures of dominance" that must be confronted on behalf of the basic human rights, as well as access to the benefits of new technologies and development opportunities, of oppressed, deprived or exploited groups, in other words, may well include traditional communal practices or tribal cultural usages as well as wielders of state power or trans-national capitalists. Here, advocacy anthropologists can learn from the example of development anthropologists such as Wright who forthrightly confront the problems of intervention within the traditional communities they strive to help as an inevitable corollary of the effort to improve their lot in relation to the national regimes and economies within which they are embedded.

One way that an anthropology of development can cross the line into advocacy is to focus on the role of documenting the limitations of existing development programs, and the ways they maintain or reinforce the social inequities they were ostensibly undertaken to ameliorate. Painter's analysis of the Bolivian program of resettlement of highland peasants in the Amazonian lowlands is an excellent example. As Painter's analysis shows, anthropologists, with their unique combination of theoretical perspectives and local knowledge, are often more qualified to provide critical evaluations of such development schemes than other social science or policy professionals. Painter's approach, in addition, exemplifies the critical sociological perspective of the best new anthropology of development: the penetration and persuasiveness of his analysis is directly related to its foundation in a recognition of the class conflicts embedded in the social situation he describes.

Winterbottom's analysis of the failure of the Tropical Forestry Action Plan exemplifies a different type of anthropological critique, one that proceeds from the opposite end of the political-economic and social hierarchy. His primary focus is at the state and international level. He points out the ways in which coordination and direction by central agencies is essential for some kinds of development-linked problems, especially those of ecological preservation. His article thus forms an ideal complement both to Wright's account of a locally focused program and to Painter's analysis of how state direction may exacerbate some of the local-level ecological problems it sets out to alleviate.

Applied anthropology and, following in its footsteps, the anthropology of development and ecological conservation policy have been theoretically humble subdisciplines, setting themselves to apply such elements of an anthropological perspective as they could, but seldom succeeding, and still more rarely aspiring to contribute in their own right to the anthropological theory of culture. It is notable how little the articles in this volume have to say about culture or, for that matter, about the potential contributions of environmentalist or developmental anthropology to anthropological theory more generally. Painter's article is a welcome exception, with his salutary reminder of how Speck's advocacy-motivated work on the utilization of environmental resources by Athabascan bands became important in the theoretical reconception of simple social and economic organization in opposition to Morgan's evolutionistic theories. The major exception in this volume, however, is Costa et al.'s article on the emergence of ecological awareness in Brazil. Costa et al. employ a class-based approach to culture as a system of conflicting formulations (in their case, "ethnoecologies") reflecting the interests of different class elements, to analyze the process by which forms of social consciousness (e.g., "ecological awareness") arise in interrelated processes of political struggle at local, national, and international levels. They thereby not only make an important contribution to the anthropology of conservation and development with their central point that the "environment," as an object of social consciousness and policy, is a cultural construct, they also provide a compelling analysis of the production of such cultural "constructs" in conflicted social and political-economic processes (a dimension too often missing from academic anthropological analyses of "cultural construction").

References Cited

Turner, Terence
  1993 From Cosmology to Ideology: Resistance, Adaptation and Social Consciousness among the Kayapo. *In* Cosmology, Values, and Inter-Ethnic Contact in South America. South American Indian Studies, 2, September. T. Turner, ed. Pp. 1–13. Bennington, VT: Bennington College.

About the Contributors

Alberto Costa is a doctoral candidate in anthropology at the University of Michigan, where he is completing a dissertation on ethnicity, race relations, Brazilian nationalism, and the economic and political significance of the ideology that identifies Brazil as a racial democracy. He has conducted long-term fieldwork in rural Brazil studying family organization, ethnic relationships, local history, and the mass media. Currently, he is working on research projects investigating "The Emergence of Ecological Awareness in Brazil" and "Strategies for Participatory Development in Northeastern Brazil."

David Gow is a development anthropologist with some 20 years of experience in various aspects of third world rural development, with a focus on agriculture, forestry, and environmental issues. During this time he has concentrated on policy formulation and planning; applied research on development issues; and project design, implementation, and management, work that has included both social and institutional analyses. More recently, he has focused on natural resource management and sustainable agriculture, environmental assessment, and resource sustainability. He has just completed a manuscript for a book that critically examines the relationship between anthropology and development in the third world from the perspective of the practicing anthropologist. An independent consultant, he lives in Washington, DC.

Conrad Phillip Kottak is professor of anthropology at the University of Michigan. He has conducted fieldwork in cultural anthropology in Brazil, Madagascar, and the United States. His general interests are in the processes by which local cultures are incorporated into larger systems. This interest links his earlier work on ecology and state formation in Africa and Madagascar to his more recent research on global change, economic development, national and international culture, and the mass media. Currently, he is directing research projects investigating "The Emergence of Ecological Awareness in Brazil," "Deforestation in relation to Variant Land-Use Patterns in Madagascar," and "Strategies for Participatory Development in Northeastern Brazil." He is a past chair of the General Anthropology Division of the American Anthropological Association (1990–1992).

Michael Painter is a senior research associate at the Institute for Development Anthropology and an adjunct associate research professor at the State University of New York at Binghamton. His general interests are in the areas of political economy and agrarian change, and the use of history in anthropological theory. He has extensive research experience on food and agriculture and land degradation in Bolivia, Peru, and Ecuador. Recent work includes a volume coedited with William H. Durham entitled *The Social Causes of Environmental Destruction in Latin America*. His current research focuses on rural production relations and patterns of capital accumulation in Bolivia, the relationships among class, ethnicity, and gender in rural social movements, and social conflicts associated with the establishment and maintenance of protected areas in Latin America.

Rosane M. Prado received her doctorate from the Post-Graduate Program in Social Anthropology of the National Museum (Federal University of Rio de Janeiro) in 1993 for a comparative study of small-town life in Brazil and the United States. Her fieldwork in Brazil has focused on issues of family and gender identities in relation to mass media. Her comparative research in Brazil and the United States has examined the contrast between rural and urban identities and the nature of small-town life. Recently, she has worked on research projects investigating "The Emergence of Ecological Awareness in Brazil" and "Strategies for Participatory Development in Northeastern Brazil."

Pamela J. Puntenney is founder and executive director of Environmental and Human Systems Management. She has served in an advisory capacity for such counties as Bolivia, Nepal, Kenya, the United States, and Korea on developing a national strategy for environmental education and has consulted with such organizations as the World Wildlife Fund, the World Bank, the U.S. Forest Service, and the World Conservation Union. As an adjunct research scientist with the Department of Anthropology at the University of Michigan, her current work focuses on issues of public choice and public responsibility in relation to the internationalization of environmental issues. She is serving as advisory to the film *Right To Hope,* which links culture, art, and the global environment as part of the 1995 celebration of the United Nations' 50th anniversary. She serves as a member of the NAPA Governing Council and is currently on the Board of the High Plains Society for Applied Anthropology. She is a past program chair for NAPA (1989–1991) and served on the American Anthropological Association Executive Program Committee (1992).

John Stiles is a doctoral candidate in anthropology at the University of Michigan. He has studied nationalism in the Caribbean and in the modern Middle East and has written on contemporary migration and the development of national identity. Currently, he is doing fieldwork on class formation, migration, and national identity in Martinique.

Roy Rappaport is the Walgreen Professor for the Study of Human Understanding at the University of Michigan, Department of Anthropology. He has conducted ecologically oriented fieldwork in New Guinea and recently worked on the social and cultural impacts of nuclear waste disposal and outer continental shelf oil drilling. He has strong interests in ritual as well as in ecology. *Pigs for the Ancestors* and *Ecology, Meaning, and Religion* are among his works. He has served as president of the American Anthropological Association (1987–1989).

Terence Turner is professor of anthropology at the University of Chicago. He received his doctorate in social anthropology from the Department of Social Relations at Harvard in 1965, focusing on the social and political organization of the Northern Kayapo of Central Brazil. He has made eight further field trips to the Kayapo, combining research, advocacy, and visual documentation and has published extensively on Kayapo social organization, myth, ritual, and history. He served as field and editing consultant for several ethnographic films on the Kayapo made by the BBC and Granada Television in the "Disappearing Worlds" series. Since 1989 he has been working with the Kayapo on a project of self-documentation of their own culture and political confrontations with Brazilians using video. He has been active in advocacy and human rights work on behalf of the Kayapo and other Amazonian peoples, in collaboration with a number of American, European, and Brazilian indigenous support organizations, and is presently a member of the Commission for Human Rights of the American Anthropological Association. He is currently working on *A Critique of Pure Culture,* a book criticizing cultural approaches in anthropology.

Robert T. Winterbottom is senior natural resources specialist with the International Resources Group, Ltd., Washington, DC. He currently serves as natural resource management advisor to the Ministry of Agriculture and Livestock of the Government of Niger, under the USAID-funded Agriculture Sector Development Grant—Phase II program in Niger, West Africa. He was formerly the director of the Forestry and Land Use Program of the World Resources Institute's Center for International Development and Environment, Washington, DC. He has a background in natural resources management, forestry, and environmental planning, with some 20 years professional experience, chiefly in the developing countries of sub-Saharan Africa.

R. Michael Wright is the president and chief executive officer for the African Wildlife Foundation. He studied law at Stanford University and was a researcher with Stanford's Center for Research in International Studies before joining The Nature Conservancy as its first director of International Programs. He served as assistant director of President Carter's Task force on Global Resources and Environment to follow up the Global 2000 report. As vice-president and general counsel with the World Wildlife Fund, he initially directed the Parks Program and eventually developed the first field-based program using conservation approaches to alleviate poverty. As senior vice-president he oversaw programs in Asia, Latin America and the Caribbean, and Africa. He has also served as a member of the U.S. delegation to the Governing Council of the United Nations Environment Programme. In 1988 he received a FUNEP 500 Award from the Friends of the United Nations Environment Programme. Works include a forthcoming book coedited with David Western entitled *Natural Connections: Perspectives on Community-Based Conservation.*

napa bulletins

Why are anthropologists joining together in local practitioner organizations? What do anthropologists in government agencies do? How does one set up and operate a research and consulting business?

These are some of the questions answered in recent issues of the NAPA Bulletin, a monograph series for practitioners in the social sciences published semiannually by the National Association for the Practice of Anthropology, a section of the American Anthropological Association.

The following issues are now available:

2 Business and Industrial Anthropology: An Overview
Marietta L. Baba
$4.50 (members), $6.00 (nonmembers)

4 Research and Consulting as a Business
Nancy Yaw Davis, Roger P. McConochie, and David R. Stevenson
$2.00 (members), $4.00 (nonmembers)

5 Mainstreaming Anthropology: Experiences in Government Employment
Karen J. Hanson, ed., John J. Conway, Jack Alexander, and H. Max Drake
$2.00 (members), $4.00 (nonmembers)

6 Bridges for Changing Times: Local Practitioner Organizations in American Anthropology
Linda A. Bennett
$2.00 (members), $4.00 (nonmembers)

7 Applied Anthropology and Public Servant: The Life and Work of Philleo Nash
Ruth H. Landman and Katherine Spencer Halpern, eds.
$2.00 (members), $4.00 (nonmembers)

8 Negotiating Ethnicity: The Impact of Anthropological Theory and Practice
Susan Emley Keefe, ed.
$2.00 (members), $4.00 (nonmembers)

9 Anthropology and Management Consulting: Forging a New Alliance
Maureen J. Giovannini and Lynne M. H. Rosansky
$6.00 (members), $7.50 (nonmembers)

10 Soundings: Rapid and Reliable Research Methods for Practicing Anthropologists
John van Willigen and Timothy L. Finan, eds.
$10.00 (members), $13.50 (nonmembers)

11 Double Vision: Anthropologists at Law
Randy Frances Kandel, ed.
$10.00 (members), $13.50 (nonmembers)

12 Electronic Technologies and Instruction: Tools, Users, and Power
Frank A. Dubinskas and James H. McDonald, eds.
$10.00 (members), $13.50 (nonmembers)

13 Race, Ethnicity, and Applied Bioanthropology
Claire C. Gordon, ed.
$10.00 (members), $13.50 (nonmembers)

14 Practicing Anthropology in Corporate America: Consulting on Organizational Culture
Ann T. Jordan, ed.
$10.00 (members), $13.50 (nonmembers)

15 Global Ecosystems: Creating Options through Anthropological Perspectives
Pamela J. Puntenney, ed.
$12.50 (members), $15.00 (nonmembers)

Please include payment, in U.S. funds, with all orders.

American Anthropological Association
4350 North Fairfax Drive, Suite 640
Arlington, VA 22203-1621

Virginia residents please include 4.5% sales tax to total.